PENGUIN HANDBOOKS

BACKPACKING ON A BUDGET

Anna Sequoia is codirector of the Climbing School of North Country Mountaineering, Inc., in Hanover, New Hampshire. A former editor and book reviewer, she has written articles for a broad range of magazines, including *New York, New Times, Viva,* and *Climbing.* She is coauthor of *The Climbers Sourcebook.* Anna Sequoia is also a photographer, with several one-woman shows to her credit. Ms. Sequoia is a member of the New Hampshire Chapter, Appalachian Mountain Club, and of the British Mountaineering Council. She is a cofounder and on the steering committee of the Coalition of New York Women Artists.

Steven Schneider is codirector of the Climbing School of North Country Mountaineering, Inc., which he founded with his sister, Anna Sequoia. He originated and taught the first technical mountaineering courses ever offered for academic credit by an East Coast college. Over the past three years alone, Mr. Schneider has been responsible for eight first ascents on Cannon Cliff, in New Hampshire, on Lenin Rock, in Washington, and on Robin Rock, in Rocky Mountain National Park, Colorado. One of his ascents on Cannon Cliff was noted in *Climbing* magazine as one of the best climbs in the eastern United States. He is also credited with the first ice ascent of Parasol Gully, Dixville Notch, New Hampshire. Mr. Schneider's articles on backpacking and climbing have appeared in *Summit, Climbing, Appalachia,* and *Viva.* He is coauthor of *The Climbers Sourcebook,* a member of the British Mountaineering Council, and former Vice Chairman for Mountaineering Activities of the New Hampshire Chapter, Appalachian Mountain Club.

BACKPACKING ON A BUDGET

**BY ANNA SEQUOIA
AND STEVEN SCHNEIDER**

PENGUIN BOOKS

Penguin Books Ltd, Harmondsworth,
Middlesex, England
Penguin Books, 625 Madison Avenue,
New York, New York 10022, U.S.A.
Penguin Books Australia Ltd, Ringwood,
Victoria, Australia
Penguin Books Canada Limited, 2801 John Street,
Markham, Ontario, Canada L3R 1B4
Penguin Books (N.Z.) Ltd, 182–190 Wairau Road,
Auckland 10, New Zealand

First published 1979

Library of Congress Cataloging in Publication Data
Sequoia, Anna.
Backpacking on a budget.
1. Backpacking—Equipment and supplies.
I. Schneider, Steven, 1951–   joint author.
II. Title.
GV199.62.S46   688.7'6'5   79–427
ISBN 0 14 046.365 8

Printed in the United States of America by
Offset Paperback Mfrs., Inc., Dallas, Pennsylvania
Set in Video Comp Eterna

——————————————————To Ghog and Booh

CONTENTS

CHAPTER 3: HOW TO BUY A NEW TENT 49

———————————————————————————————— **About Tents 49**

CHAPTER 4: HOW TO BUY A NEW SLEEPING BAG 66

——————————————————————————— **About Sleeping Bags 66**

ACKNOWLEDGMENTS

This book could not have been written without the cooperation of the following people: Tim Arnstein, Precise Imports; Ray Arvio; Mary Bender; James F. Booker, E.I. du Pont de Nemours, Inc.; Norman Brilliant; George and Ida Cohen; Rebecca Dominguez-Blum; Anna Falco; Joan Ford, The Ashflash Corporation; Barry Goldensohn, Dean, Hampshire College; Anton Goldman; Nancy Grimes, Banana Equipment; Anne Rebeck Hajny; Dick Hathaway, Goddard College; Ned Hitchcock, Goddard College; Larry Horton, Rivendell Mountain Works; Donna Hunter; Chuck Kennedy, Arcadia Transit Authority; Sally Lamb; Rosalie Laughlin; Jerry Lavin, E. I. du Pont de Nemours, Inc.; Mike Lowe, Lowe Alpine Systems; Bill Mason; Doug "Hamish" McBride; Katie McMullen, Coleman Corporation; Kristine H. Meyer, Damart, Inc.; Hope Moskowitz; Vincent Natelli, United States Postal Service; Alan Ravage; Eric Reynolds, Marmot Mountain Works; Patty Rosenberg; Hal Schneider; Nancy Jane Schneider; Herbert H. Sheinbaum, Sheinbaum, Inc.; Sam Smith; Joseph Tanner, W. L. Gore & Associates; Greg Wehrli; Fred Williams, Moonstone Mountaineering; Ron Zimmerman, Early Winters, Ltd.; Sue Zuckerman.

INTRODUCTION — INCLUDING THE SEVEN KEYS TO BUDGET BUYING

They were the biggest trees we had seen up to that point—acres and acres of huge Ponderosa pine. The only sound when we woke was the stream rushing in its deep gully down from the Anderson Glacier to the sea. The water was alive with trout. There is a curious quality to the light in a great forest on a morning like that: diffuse, broken into luminous patches by the high, thick branches or concentrated into long, dense, oblique beams, like afternoon light in a cathedral. It was the most peaceful and comforting awakening either of us could imagine—and what takes us back, time after time, to the forests of the Pacific Northwest.

Backpacking, for us, has been full of glorious moments: emerging from the Mount Willard Trail to a sudden view of the stark cliffs and gullies of Crawford Notch in New Hampshire's White Mountain National Forest; sitting in a cold, damp cave in Wyoming, brewing tea, watching the mist rise straight up the vertical Dike Route of the Middle Teton; or just resting at the edge of the sea in Acadia National Park in Maine, calmed by the movement of the tide.

The irony is that over the eighteen years that we've been backpacking, the cost of going out into the natural world has quadrupled. Over the past twelve years alone, the cost of outfitting even a sometime backpacker has risen dramatically. For example, in 1967 a Camp Trails magnesium frame with its best pack bag cost $26; today that same outfit costs about $92. A down sleeping bag that cost $50 now sells for about $140. Sturdy hiking boots that sold for between $16 and $20 now cost between $68 and $82. An even more stunning example: a good compass that retailed in 1967 for $4.00 can now command as much as $26.

Even the cost of necessary "incidentals" has risen almost absurdly. Heavy wool socks that sold for under $2.00 now sell for $9.00. An Army surplus canteen that was commonly sold for a quarter now costs $2.00. A simple Swiss pocketknife that used to cost between $3.50 and $5.00 today costs $13 to $25.

This book is designed to help you beat the high cost of wilderness backpacking. If you're a novice, you'll learn what equipment you need—and what you don't. If you're an experienced backpacker, you'll find this book chock full of new equipment information—and methods of budget buying that may never have occurred to you before. And if you're one of the thousands of backpackers who take your family into this country's forests and wilderness areas with you, you'll be able to save hundreds of dollars on the cost of equipping your backpacking clan.

What this book is, really, is a consumer's guide for backpackers. If you follow our advice, you should be able to save 30 to 50 percent on all your backpacking needs—from backpacks, parkas, and tents to freeze-dried food.

Almost all the information in this book is available for the first time. No other book gives you, for example, information on forming an equipment-buying cooperative; or advice on where and how to buy secondhand gear; or how to beat inflation by buying gear now and putting it away for later use.

We have researched and compiled, for the first time anywhere, a comprehensive, state-by-state guide to stores that discount name-brand equipment, sell manufacturers' overruns and seconds, rent backpacking equipment, and offer discounts to nonprofit organizations. Included also, for the first time in any book, are the dates these stores hold their annual and semiannual sales.

Here, too, is a comprehensive first look at a fabric that is revolutionizing the outdoor equipment industry, Gore-Tex®; and the most complete directory yet published of manufacturers and mail-order sources of outdoor equipment kits.

We have successfully used all of the techniques for saving money on outdoor gear that we have included in this book. All of them work.

──────────── **The Seven Keys to Budget Buying**

Each bit of buying advice that follows has, over the years, saved us hundreds of dollars on our outdoor gear. These approaches have worked for our friends and our clients, and they can work for you too.

1. Read the catalogs and know the prices. The more you know about backpacking and how much it costs, the easier it'll be for you to recognize a bargain when you see one. We suggest that you familiarize yourself with all of the following catalogs: the Eastern Mountain Sports catalog (send $1.00 to EMS, Vose Farm Road, Peterborough, New Hampshire 03458); the Recreational Equipment, Inc., catalog (send $1.00 to REI, P.O. Box C-88125, Seattle, Washington 98188); the Holubar Mountaineering Ltd. catalog (free from Holubar, P.O. Box 7, Boulder, Colorado 80306); the North Face catalog (free from The North Face, 1234 Fifth Street, Berkeley, California 94710). One reason you'll be studying these catalogs is to see what items you want and need. But don't impulsively send away for them at full retail price. Our motto is: *Never pay full retail price.* It's just not necessary. Once you're clear about what you need, you'll be able to plan ahead to buy it for less. If, for example, you need a new down-filled vest, you'll be able to consult Chapter 10 of this book—and find out when the store in your area that carries that item will be having a sale. Or you may decide to buy one secondhand, or to look for a discontinued style, or a manufacturer's second. What we're saying is that once you know what you need, you can in good time examine all the options for buying it at less than full retail price if you plan ahead.

2. *Buy just what you need.* Once you've received all of the catalogs we've mentioned, you must remind yourself that catalogs are designed to make you part with your money. So are most backpacking equipment retail stores. In most cases, you don't need the same fancy equipment that was used on the latest K2 or Everest expedition; don't fall prey to the snob appeal of advertising pitches for most equipment. Let's take an example. Most people spend far too much on hiking boots. Most beginning hikers will be going out on one- to four-day hikes that rarely cover more than twenty miles—and they don't need $80 Fabiano Mountain Boots. Nor do they need Norwegian welt construction or Goodyear welt construction to get good footing and traction. This past summer two of our friends hiked up Long Peak in Colorado wearing just Nike Waffle Trainers (essentially, fancy sneakers). These provided good support, good traction, and turned out to be excellent lightweight trail shoes. We're not recommending, of course, that you get yourself a pair of sneakers and go off hiking in them. But there is a lesson to be learned here: In past summers, the same two friends wouldn't have considered hiking in anything less than their Galibier Superguides.

There does seem to be a trend toward lightweight trail shoes that supply good ankle support without the burden of extra poundage —the Danner 64/90, for example, or the Fabiano Madre 90. Most people who went hiking and climbing in the Grand Tetons this year were wearing very light shoes which could be used for the trail, rock climbing, and light-snow climbing.

Personally, we think this is a trend that should be encouraged. We've seen the horrid toll five- and six-pound hiking boots with deep Vibram soles have been wreaking on our mountain trails over the past few years. We've seen turf pulled up, moss destroyed, and very heavy erosion—all caused by people who aren't even the types to leave litter behind them. It's unconscious destruction.

But there is another great benefit in this trend toward lighter footgear: lighter footgear costs less.

The same principles that apply to buying boots apply to backpacks and to most other equipment. Do you really need an $85 Kelty D-4? Will a $55 REI Panel Pack Cruiser do? It holds just as much, and will last just as long.

3. Rent what you can, instead of buying. Say you're going winter backpacking three times this winter. Does it make sense to go out and buy a winterweight sleeping bag, a $50 winter stove, and an ice ax? Probably not. Chapter 10 will tell you where you can rent this kind of equipment at very reasonable prices. Renting a winterweight bag for four days, for example, shouldn't cost more than $14. Renting a winter stove for that period may cost up to $8.00. An ice ax rents for about $5.00 for a week. Eventually, if you find that you're really committed to winter backpacking, you can always plan ahead—and shop thoroughly before you buy.

If you're a beginning backpacker, renting your equipment is one of the best ways to try out different kinds of gear (you'll also want to borrow as much gear as possible from friends—for the same reason). Renting will give you the time you need to carefully think over what kind of equipment you really want—and to save up for it, and for fall sales.

Rental departments are, in most areas of the country, getting more and more extensive. At some stores, like the Boulder Mountaineer in Boulder, Colorado, for example, you can rent practically everything you'll ever need—including boots, technical climbing hardware, haulsacks, lanterns, and compasses. That's just for starters.

4. Shop at sales. One of the most frustrating things about buying new equipment is that you can go into a store, buy a new tent or some other major piece of gear, and then two weeks later see the same item on sale for $30 to $60 less than you paid. We know how hard it is, once you've made up your mind what equipment you want, not to run right out and pay full retail price. But if you want to save money, and have more money for all the rest of the equipment you'll need, you must plan ahead. We advise you to wait for the sales. In the meantime, if you absolutely must have a piece of equipment by the weekend, borrow or rent it; in most cases, even paying an occasional rental fee, you'll still wind up paying less for your gear if you wait for the sales.

Even if you're *contemplating* buying a new tent or some other major item, get yourself on catalog and retail store mailing lists. If they have a big sale coming up, they'll let you know. Also, watch the newspapers. Most stores have huge annual or semiannual sales. But many also have good special one- to three-day sales, often around holidays like Washington's Birthday.

Once you've learned what's on the market, you'll be in a position to narrow your choice down to two or three brands or models. Chances are, if you're among the very first on line the morning the sale begins, you'll find at least one of the models you want.

Using these simple techniques ourselves over the years, we've picked up very substantial buys. Recently, we were able to purchase a North Face St. Elias tent at EMS in Boston for under $100—60% off list price. We were also able to pick up an Alpine Designs Logan Tent at an EMS Ardsley store sale for $105—at least 50% off list price. In both cases, it was just a question of knowing what we wanted, waiting for the sales, and being persistent. Why contribute an extra hundred dollars revenue to EMS or Holubar when you can use that hundred dollars yourself?

5. Buy manufacturers' overruns and seconds. When any piece of equipment is produced, a manufacturer will make more than is necessary to fill its orders. At the end of each fiscal year, these excess items are offered to retailers at considerable savings. Most retailers will pass these savings on to you at special annual clearance sales or at special sales throughout the year.

Seconds are items that have been soiled or slightly damaged during the manufacturing process. Manufacturers themselves weed out the severely damaged merchandise, so what reaches your retail store is generally in very good condition. These seconds very rarely affect the quality of the product. Savings on seconds of sleeping bags, for example, will vary from about 15 percent to 40 percent off list price. Considering how much a sleeping bag costs today, these represent very substantial savings.

The guide in Chapter 10 will direct you to the stores near you that regularly carry overruns and seconds. It may take a few visits to these stores before you happen on one of the sleeping bags or other items that interest you. If you're persistent, you're going to find one.

If you're looking for a second of a specific brand, like Sierra Designs, try the Sierra Designs' retail store first; they're the most likely to have the better seconds, and to have them most often.

And don't neglect factory outlet stores. Invariably, where there's an equipment factory, especially a boot factory, there is a factory outlet store. One of our favorites for years has been the Dunham Factory Outlet Store in Brattleboro, Vermont. Another favorite of ours is the Eureka Tent and Awning Company Factory Outlet Store in Binghamton, New York. The Eureka store offers 30 to 60 percent off on the entire Eureka Tent line.

6. *Shop at clearance centers and bargain basements.* Bargain basements and clearance centers are the Mecca of the bargain-seeking outdoorperson. Chapter 10 will guide you to stores that have bargain basements and clearance centers. These clearance centers invariably contain one or more tents, and you can expect a discount of 20 to 40 percent off manufacturers' suggested list price. Because we run a professional mountain guide service, we sometimes need several new tents at a time—and we've done very well at bargain basements and clearance centers, sometimes getting tents at less than what they'd cost us wholesale. Within the past couple of years, for example, we were able to acquire four North Face Morning Glory tents this way for $160 each—a saving of $200 per tent!

Clearance centers and bargain basements are also great places to look for boots; there always seem to be some reasonably good boots put out for clearance. Three years ago, for example, Anna found her Galibier Makalu Hivernales (double boots) in the clearance section of Skimeister Sport Shop in North Woodstock, New Hampshire. Because we bought these off-season, and because there aren't that many people who need double boots in size 6, we were able to get these excellent high-altitude mountaineering boots at $55 less than their list price. In the three years since we bought the boots, the price of Hivernales has gone up so much that these boots are now worth twice what Anna paid for them.

We've also found *fantastic* buys on sleeping bags in these places. This past winter, Steve found two Snow Lion overbags for $40 each—half their regular price. We also found two Alpine Designs High Loft Sleepers for clients of ours this winter for $67.50 each—$30 less than manufacturers' suggested list price. These bags were reduced and put in the clearance section because they were discontinued models; but they're both excellent bags.

Bargain basements and clearance centers really can be treasure troves; we recommend that you visit them regularly. It helps to get to know the salespeople too. That way, if they know you, and know you're looking for some particular item, they'll put it away for you.

7. Buy second-hand equipment.

MOUNTAIN TENT, 2 man, Gerry Southface, excellent cond., includes poles, stakes, fly and shock cords. Approx. 6 lbs. About $130 new, sell for $75.

SLEEPING BAG. EMS Franconia, $50; North Face Mountain TENT, $100. Both 1 year old, in excellent cond.

SKI MOUNTAINEERING BOOTS, Molitor 11½ N. Superb, absolutely rigid, double insulated boots for ice climbing or skiing. Very good cond. Cost $205 with boot trees, will sell for $45.

DOWN BAG, Alpine Design, large, overlapping "V" tube construction, rated to −20°; TENT, Alpine Design 2-man mountain, lightweight, can be used year-round. Both in excellent condition. $75 each.

All of the ads above appeared in a single April issue of the *AMC Times*, the newsletter of the Appalachian Mountain Club. Almost any bulletin board in the larger backpacking retail shops contains many similar announcements—and potential bargains. Does it make sense to buy brand-new equipment when you can get perfectly good equipment at these kinds of prices?

Chapter 7 tells you everything you need to know about buying used equipment: how to find it; what to look for when you examine it; and how to pay less than the asking price.

And one last word about outdoor chic: The more vintage your equipment is, the longer you've been around. If you have an old Kelty packbag with the old design label, you'll get more nods and acceptance from old pros than if you show up on the trail with a brand new red Jansport D3. So—do you really need a *new* soft pack or pack and frame? Give it some thought.

1

HOW TO BUY A NEW BACKPACK

About Frame Packs

There are many different kinds of frame backpacks. The simplest is a wooden ladder-like structure to which shoulder straps have been added. Canvas or canvas-like material is wrapped around the bottom three-fourths of the ladder to provide rather primitive cushioning. A bag or box is then lashed onto this frame. This type of frame is still very popular in certain areas of the United States: "Hut boys" (both male and female) for the Appalachian Mountain Club still carry 150-pound loads using this system.

The wooden pack frame, however, is an anachronism. Most external frame packs are now made from aluminum tubing, magnesium tubing, or molded plastic tubing. These materials are lighter, more durable, and provide a more comfortable fit.

Most backpacks produced for the American market feature a shoulder band, padded hip belt, and padded shoulder straps. Attachment of the packbag to the frame is either through rivets or through studs that pass through the backpack and the frame. Some backpacks are attached via a system of buckles.

HOW TO GET THE RIGHT FIT

It doesn't matter how much you save on a backpack if it doesn't fit. The distance between the waist belt and the shoulder straps should equal the distance between your waist and your shoulder. You can have an inch leeway either way. Any greater variance will cause trouble.

If you're going to be using the backpack during winter, the length of the frame should be one inch longer between the waist belt and the shoulder strap than the measurement between your own waist and shoulder; this will allow for proper fit when you're wearing bulky clothing.

A fully padded wraparound waist belt will make your backpack much easier to carry. But if you see a good frame at a good price without the fully padded waist belt, buy it. Don't worry about the missing padded waist belt—you can add one later.

Something you must look for in any frame pack is adjustable shoulder straps. Many inexpensive frame packs don't have them. But if you are handy and can borrow a variable speed drill, installing your own adjustable shoulder straps isn't beyond your reach. Instructions for installing shoulder straps are included in the package in which they come.

HOW TO DECIDE UPON THE RIGHT CAPACITY PACKBAG

First, have some idea of where you're going to go backpacking —and for how long. If you're a weekend backpacker, a mid-size packbag (2000 cubic inches) with separate lash strap attachments is more than adequate. If your trips take you out for a week or more, or if you're a winter backpacker, you'll need a packbag of about 4000 cubic inches. Some backpacks with this volume are the Kelty Serac, the EMS Expedition, the REI Expedition Bag, and the Adventure-16 Expedition Bag.

A good packbag will offer one main compartment (usually top-loading), or a divided bag with access to the bottom compartment through an outside zipper, or it will have an outside zipper access with no top flap. In rainy conditions, it is desirable to have a packbag made of waterproof fabric. This will avoid any problem of leakage into the pack. It also eliminates the necessity of carrying a specialized rain cover.

Some excellent quality bags, though, are not waterproof (for example, Kelty packbags). You can remedy this situation by making your own waterproof cover from waterproof nylon sold by the yard, and in doing so add less than $5.00 to the initial cost of the bag.

HOW IMPORTANT ARE SIDE POCKETS?

Side pockets are a desirable characteristic because they allow you to pack socks, film, a canteen, trail food, mosquito repellent, and raingear within easy reach. But if you find a bargain bag without side pockets, don't be too concerned; you can buy side pockets for many bags, and of course you can make your own.

Adventure-16 Pack and Frame
(Photo courtesy of Adventure-16)

HOW IMPORTANT ARE LEATHER PATCHES?

Leather patches are useful for carrying crampons or lashing incidentals to the pack, but you don't really need them. If you want them, however, you can buy them at any large outdoor equipment store and sew them on yourself. They generally cost 80¢ to $1.50 per patch.

PADDED HIP BELTS—THE ONE NECESSITY

Most backpacks today come with at least some padding attached directly to the pack frame. This provides you with the cushioning needed for moderate loads. If you intend to carry more, Camp Trails, Kelty, and Adventure-16 produce a fully padded wraparound hip and back belt that will give you miles of comfortable walking. The average cost for one of these belts ranges around $10.00 to $20.00. Naturally, if you get them at semiannual clearance time, you'll pay less.

WHAT TO LOOK FOR IN STITCHING

Check to see that your bag is evenly stitched with nylon thread. If the stitching is not even, or if cotton thread is used, your bag can weaken and tear at critical stress points. If the stitches are too far apart, this will cause a weak spot in the construction of the bag. But this does not mean that you couldn't purchase a badly sewn bag and reinforce the stitching yourself.

HOW IMPORTANT ARE FRAME EXTENSIONS?

The advantage of a frame extension is that it provides you with additional lashing points for securing tents, sleeping bags, and miscellaneous gear too bulky to fit inside the packbag. Frame extensions can be added to most pack frames. Bring your frame into your camping equipment retailer in order to assure proper fit. A frame extension is a *necessity* if you plan to go out during the winter. Extensions are not expensive, and they add real versatility to your pack.

SOME GOOD VALUES FOR THE MONEY/FRAME PACKS

A good value for the money has most or all of the features mentioned in this chapter, doesn't strain your budget, and offers more features and better workmanship than comparably priced packs. Some companies, like Eastern Mountain Sports, are aware that not all of their customers can afford their most expensive packs —so they offer one or two packs that are much more modestly priced. These packs, like the EMS Backpacker, for example, which we mention below, have a packbag that is made from fewer pieces of cloth, involves less stitching, and has fewer outside pockets. All of these things save the manufacturer money, particularly in labor costs, and these savings are passed on to you. Try to get each company's full equipment catalog; but look toward their least expensive items.

Orion Pack and Frame
(Photo courtesy of Camp Trails Company)

Astute timing and shopping, as we pointed out in our introduction, are important too. A manufacturers' second of, say, a Jansport D-3, can be a better value than buying the bottom of their line.

The frame packs that we list below are all very good values for what they cost.

Kelty Basic Pack

This is a very simple top-loading packbag, with a large single compartment and a good sturdy frame. The bag has two large outside pockets. A padded hip belt is optional and will run about $8.00 extra. It's the most inexpensive frame pack in the Kelty line, averaging about 2021 cubic inches. A great buy at less than $65.

REI* Cruiser Bag and Frame

This large, single-compartment bag with five side pockets also has a fully padded wraparound hip belt. It's strong, durable, waterproof, and is a very good lightweight recreational backpack. It should cost less than $60.

EMS Backpacker

A single-compartment (2300 cubic inches) bag with three large outside pockets, the EMS Backpacker has a fully adjustable frame. Other features of this waterproof bag are a padded hip belt and a mesh back band. It is priced at under $60.

Camp Trails Adjustable 1 Backpack

This inexpensive (just over $50), sturdily built packbag and frame comes with two side pockets and an adjustable padded hip belt. An excellent feature is the shoulder strap that enables both a tall person and a shorter person in one family to use the same bag. The shoulder strap adjustment is achieved by moving the shoulder strap connection points to a higher or lower position on the frame. This bag is waterproof too.

Himalayan System 9 Weekend Combination

The System 9 has one main compartment, four outside pockets, and a separate map pocket. It is waterproof and has a 2100-cubic inch capacity. This is an excellent starter bag for those just getting involved in backpacking and costs just over $50. It comes with a fully padded wraparound hip belt.

Coleman Peak One

This bag was introduced to the American market in the fall of 1977. It features a fully adjustable pack frame that can be used by someone who is 5 feet 5 inches or 6 feet 3 inches—giving the bag great and unusual versatility, especially for families that share equipment. The packbag itself is very roomy, over 4000 cubic inches, to accommodate winter loads. The pack is sold with a matching stuff sack and an extra storage ditty bag which fits directly onto the top of the frame—a very convenient place for your camera equipment. At around $50, this bag is a very good buy indeed.

---

*Recreational Supply and Equipment of Seattle, Washington, is the oldest backpacking and mountaineering cooperative in the United States. In 1978, it celebrated its fortieth birthday. REI manufactures much of its own gear, a good deal of which is cheaper than similar gear from other manufacturers. This gear is listed in their annual catalog and supplementary mailings. You can join the co-op for a one-time fee of $2.00. Every member receives a 10% rebate on all REI equipment purchased during the year. This rebate can be applied toward other equipment, or returned to you in cash. Their address is: Recreational Supply and Equipment, Inc., 1525 11th Avenue, Seattle, Washington 98122.

Almost every company that produces backpacks produces a line of soft packs (also known as rucksacks). They come in two categories: frameless, and internal frame construction.

There are two types of frameless packs: a hanging pack which relies on shoulder straps to keep the bag on your back; and the contour pack, which is cut in a similar shape as your back. With careful packing, the latter can provide the most comfortable means of carrying any gear. We have carried up to 60 pounds in our contour packs (one Cyclops, and a Giant Jensen by Rivendell) and have been perfectly happy with them.

An internal frame system can consist of two soft aluminum stays which can be bent manually to contour to your back (for example,

Coleman Peak One Pack and Frame
(Photo courtesy of the Coleman Company)

the Alpine Designs Eiger Tour Pack). Another internal frame system uses a single piece of molded fiberglass (the North Face Wrapac). Others use crossed and x-shaped unremovable 1½-inch unbendable aluminum upright stays (as in the Hine/Snowbridge Serex). Still others use sewn-in foam padding for support and comfort (Wilderness Experience, and Kelty).

A soft pack is particularly useful for day hikes when you may want to carry just raingear, lunch, compass, and sweater. For this kind of load, no frame is necessary, as no great stress is being put on the shoulders or the back. But there is no reason why you couldn't use a soft pack for longer hikes too; you just need one with appropriately larger capacity.

Small soft packs can be purchased at outdoor equipment suppliers, discount department stores, or military surplus stores. They are made of cotton duck, canvas, or nylon. The cheapest way to get a small soft pack other than at a sale is to make one yourself. It is not particularly complicated and is not beyond the skill of a begin-

Giant Jensen Pack
(Photo courtesy of Rivendell Mountain Works)

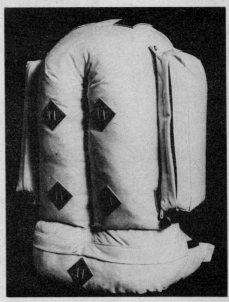

The Hine/Snowbridge Serex
(Photo courtesy of Hine/Snowbridge)

The Country Ways Kit Pack
(Photo courtesy of Country Ways Kits)

ning sewer. Excellent soft pack kits are produced by Altra, Inc., Carikit Outdoor Equipment Kits, Country Ways, Eastern Mountain Sports, Frostline, and several other kit companies. These kits can save you more than half of what you'd pay for a presewn soft pack.

HOW TO CHECK FOR THE BEST CONSTRUCTION

Check to see how securely the shoulder straps are attached to the neck of the pack. You don't want a shoulder strap ripping or tearing at either the waist or the neck.

Check to see that the tie-down straps or closures are well constructed and attached strongly.

Check all seams for weak stitching and loose ends. If the bag is improperly sewn, look for another bag (or if it's cheap enough, reinforce the stitching yourself).

Also, be sure the zipper is well-sewn to the pack and that the zipper opens and closes easily.

Don't be influenced by the amount of leather patches attached to the bag; you can add these yourself for less than $2.00 each.

SOME GOOD VALUES FOR THE MONEY/SOFT PACKS

Chuck Roast Alpine Pack

This inexpensive pack is made of water-repellent cotton canvas. It has a 7-inch extension at the top of the bag for increased capacity. It has padded shoulder straps, and even at manufacturer's suggested list price (just over $40) is a bargain for the money. The bag itself has a capacity of 1700 cubic inches. It is available through Eastern Mountain Sports, or directly from Chuck Roast Equipment, North Conway, New Hampshire.

McKenzie Pack

Constructed of 18-ounce linen canvas, the McKenzie Pack has a ⅘-oil-tanned leather bottom, padded shoulder straps, and an ice ax loop. Capacity is 1500 cubic inches. This nicely constructed, durable, and very rugged soft pack costs under $50 and is available through Great World, Inc., West Simsbury, Connecticut.

REI Summit II Pack

This two-compartment bag in a teardrop shape for climbing or hiking is made of coated nylon pack cloth with an all-leather bottom. It also features padded shoulder straps. At just over $20, this bag is a very good value indeed.

Wilderness Experience Kletter Sack

A handsome bag made of cordura nylon, the back has ½-inch foam padding. There's a top flap for storage of extra gear, fully padded shoulder straps, and lots of leather patches for additional gear. It costs under $35.

Camp Ways Rambler

A large teardrop-shaped bag, the Rambler is made of 8-ounce coated nylon. It offers the backpacker two compartments, padded shoulder straps, and a padded waistband. Another good value for the money at around $30.

Trailwise Ski Tourer

Made of waterproof, heavy-duty cordura nylon, the Tourer has padded shoulder straps and waistband, and quick-release buckles. This teardrop design pack is well-made and functional. It costs less than $45.

2

HOW TO BUY NEW BACKPACKING BOOTS

——————————————————————————— **About Boots**

Most backpackers pay too much for their boots. The most common (and expensive) mistake is to buy a boot designed for use in conjunction with crampons when a nice lightweight trail shoe will do more than an adequate job.

There are over fifty different brands and models of boots on the market right now. Most boots available in the United States are constructed in the following four ways: with the Norwegian welt, the Goodyear welt, the cemented welt, and the injected welt.

The following discussion of boots, with the exception of the description of Goodyear welts, is from an excellent pamphlet produced by the Raichle Company. Read the following pages carefully before shopping for boots.

HOW BOOTS ARE CLASSIFIED

Nothing is so critical to walking comfort as hiking boots that fit well. But with dozens of manufacturers and hundreds of models, how can you find the boot that is right for you?

The first step in taking the mystery out of boots is to understand that *all boots are classified according to intended use.* Boots are divided into four main categories: (1) trail shoes, (2) hiking boots, (3) kletterschues and climbing boots, and (4) mountaineering boots. Each type is designed and constructed for a *specific purpose.* By determining which category fits your needs, you can quickly find the boot that is right for you.

Trail shoes (1), as the name implies, are *more shoe than boot.* They are meant for gentle forest trails where the footing is relatively easy, the load (if any) is light, and strong support is not required. Trail shoes vary considerably in quality—from "campus boots" and "waffle stompers" to very respectable lightweight boots. Trail shoes enjoy considerable popularity because they are extremely light, require little breaking in, and fit the fashion of the day. They range in price from $20 to almost $40. The best trail shoes are well-suited to light hiking at best.

Hiking boots (2) are substantially better for walking over the land than the best trail shoes. Designed to support and protect the foot, they are built for use on rather rugged, rocky trails by hikers who may be carrying substantial loads. Constructed in a totally different manner from trail shoes, they are tougher, heavier, and stiffer. They require a breaking-in period and usually cost from $45 to $75. The hiking boot comes closest to being an all-purpose boot and is the most popular of the categories. Hiking boots have a sturdy design, use high-quality materials, and provide dependable protection and support for the foot under a variety of outdoor conditions.

Climbing boots and kletterschues (3) (literally climbing shoes) have one important feature in common—both are designed for climbing *beyond the trail.* Except for this feature, they are quite different. Climbing boots are built for cross-country travel on rock and snow where the foot must be protected against extreme punishment and stress. In addition, the climbing boot must adequately support the foot when you cling to the edge of a ledge. Climbing boots are big, stiff, tough and usually cost $50–$70. Kletterschues, on the other hand, are thin, tight-fitting unlined shoes with laterally rigid, close-cropped soles. They are designed to place the climber's foot as close to the rock as possible on a sole suitably stiff for edging. Kletterschues, though they cost only about $18–$35, are so specialized (and afford so little support or protection) that they are used specifically for friction climbing.

Climbing boots are as heavy as kletterschues are light, often weighing 5–7 pounds. Although currently in vogue among those who feel "the bigger the boot the higher the fashion," comparatively few hikers can rationally justify these stiff, heavy boots—especially when you remember that "one pound on the foot is like five on the back." Whatever the classification, you must be a weight-watcher to avoid ending up with too much boot for your intended use.

Mountaineering boots (4) are like climbing boots. Large and stiff, heavy and hard to break in, they are designed for *serious, rigorous, expeditionary use* on ice, rock, and snow, where insulation, protection, and performance are vital to both success and survival. Price and weight discourage most fashion seekers, and comparatively few of these big boots are purchased accidentally. Cost: $75 to well over $100.

In summary, then: Trail shoes are made for light loads on good trails. Hiking boots are for any load on the toughest existing trails. Kletterschues are for technical climbing on rock. Climbing boots are for traveling cross-country. And mountaineering boots are for the ultimate in expeditionary climbing. Choosing the right category is your first step in the right boot for you.

THE BASICS OF BOOT CONSTRUCTION

After narrowing the search on the basis of intended use, you should know and understand materials and construction in order to select a model and brand. Boots are most easily classified by the manner in which the soles are fastened to the upper boot. The four main methods are: (1) cemented, (2) inside stitched, (3) welted construction, and (4) injection molded. Although the lug soles on all boots are fastened with adhesives, the distinctions between the four different methods are quite important.

Cemented boots have the upper leather folded under a light inner sole and a one piece rubber lug sole is then glued on. There is no stitching, no midsole, very little support or protection, and the boots cannot be resoled. This is the most inexpensive type of construction and is found only in the trail shoe category. Cemented boots can be recognized by the absence of stitching and the lack of a midsole. (The midsole is the layer of leather or rubber between the upper and the outer sole.)

Inside stitched boots have the upper turned under and sandwiched between a stout inner sole and midsole. The "sandwich" is fastened with a double row of lock stitching concealed within the boot and is thus protected from abrasion, moisture, and drying. The soles can be closely cropped (trimmed) so that hiking boots can double for rock climbing. On inside stitched boots, both midsole and outer sole are easily replaced.

The shortcomings with this simple method of construction are the failure of the sole to flex easily (the inner and midsole are stitched tightly together which stiffens the entire sole) and to mold itself to the bottom of the foot. Soles tend to be stiff and uppers do not always conform to the shape of the foot. Inside stitching is found on many trail shoes, nearly all kletterschues and some hiking boots, but never on climbing or mountaineering boots. Inside stitching is recognized by the absence of outside (visible) stitching, the presence of a midsole and a double row of stitches (usually covered by a synthetic sock liner) in the inner sole.

The oldest and most common type of boot construction involves outside stitching and employs the term "welt." There are several types of welt construction.

The most popular type of boot construction is the *Norwegian welt,* proven over the years to be both durable and functional. Welted boots are recognized by one or two (sometimes three) lines of stitching along the narrow shelf where the upper meets the sole. The first line slants inward to fasten the upper boot to a lip on the bottom of the inner sole. The second runs straight down to fasten the upper, or welt, directly to the midsole. The welt itself (which is omitted on a majority of welted boots) is a narrow, somewhat vulnerable strip of leather that runs around the boot on top of the midsole and hides the first line of stitching. The third line of stitching is found only on true mountaineering boots.

In a welted boot, the filled space between the inner sole and midsole permits easier flexing of the sole, which in turn reduces foot fatigue. The welt also enables the inner sole to conform comfortably and precisely to the shape of the foot because the last remains in the boot until the very end of construction. Welted construction is the most desirable way of making quality boots and is found in nearly all the climbing and mountaineering models made, as well as a majority of hiking boots.

Injection molding is a relatively new process. In attaching the upper to the sole, molten neoprene, applied under pressure, takes the place of stitching and cement. There are two main types of injection molded boots. The first finds the lug sole is molded directly to the upper. This type, found only in trail shoes, is comparable to the inexpensive cemented process. The boots are not easily resoleable. The second type finds a separate midsole, usually of rubber, sandwiched between the molded section and the outer sole, making the boot easily resoleable. This type is used mainly for mountaineering boots.

Injection-molded soles adhere well only to boots which use dry (low grease content) leather, which must be silicone-impregnated for high water repellency. Consequently, there is little comfort and breathability which is associated with leather insoles and midsoles. Injection molding offers economy and light weight, and has some advantages over stitching; i.e., no wear and rotting of the thread; no penetration of water through the needle holes; and total sole stiffness can be obtained for technical rock climbing.

Next in importance is the quality of leather used in boot uppers. The tanned cowhide most often used can be classed as either a "split" or "top grain". When the hide comes off the cow it is too thick for boot uppers and consequently must be split into sheets of varying thickness. The outside or top layer is called "top grain." It is ideal for boot uppers because of its tough, flexible oiliness and natural resistance to water and moisture. All other layers are called "splits." Though comparatively inexpensive, "split" leather is inferior for boot uppers because it tends to stretch and is difficult to waterproof adequately. A "rough out" leather may be reversed grain (smooth side in, rough side out) or it may merely be a sueded split (which is not as good). The difference is considerable and important.

The hide of the female animal is grouped by age: calfskin, heifer, cowhide. The hide of the male animal is classified according to its condition. Leather is usually classified in three price categories: Grade A, B, C. Grade A is the most expensive because of its spotless and flawless structure. In addition, every hide is divided into surface groupings according to quality. The best portion of the animal is the Croupon area. Leather from other parts of the hide such as the stomach and leg is definitely inferior.

You should ask the salesman (or manufacturer) for the leather specifications to make the distinction. Most salesmen will not hesitate to brag about the leather used in the boots they carry—if it is truly superior top grain.

Since seams in the leather upper boot are vulnerable to breakage and leaking, the general rule is "the fewer the better." The one-piece upper is ideal, but it wastes leather and you must be willing to pay a premium for seamless uppers or settle for inferior quality leather. All boots (except kletterschues and some trail shoes) are lined and reinforced in a variety of ways, using a variety of materials. The better-constructed the boot, the better the foot protection and insulation provided. All quality boots should have "heel counters" that cup the heel in a grip that helps anchor your foot to the sole and prevents excessive heel lifting. Also there should be a boxed or stiffened toe. High-quality boots provide "shanks" in the sole to support the arch and protect the instep. The "shanks" are either of steel, laminated wood, or very stiff plastic. Most boots provide foam rubber padding throughout the ankle area as well as a roll of foam that forms a Scree Guard or collar around the top of the boot. The best boots are carefully designed to provide softness and flexibility where possible, limiting stiffeners and reinforcement only to places where they are needed for support and protection. Inexpensive boots often attempt to achieve protection and support by using stiff materials throughout, making the boots heavy, uncomfortable and hard to break in. The midsole above the lug sole (not found in cemented boots) provides stiffness, support, and comfort. Leather midsoles help shape the boot to the foot and provide breathability and absorbtion of perspiration that promotes a cooler walk. Nevertheless, rubber midsoles are increasing in popularity with some manufacturers as the price of leather steadily rises.

There is general agreement on outer soles—a lug pattern of high-carbon neoprene material is unexcelled. *Vibram* is the brand currently in vogue and it is excellent. The softer, lighter, shallower tread of Vibram Roccia (model name) is suitable for some boots, but the firmer, deeper, heavier Montagna provides more wear and protection.

Goodyear welt construction looks similar to Norwegian welt construction. The uppers of the boots are stitched to the innersole, and then stitched to the midsole. All of this stitching appears on the outside of the boot and is very vulnerable to water and breakage. Goodyear welt construction is far less durable than the Norwegian welt and is much more difficult to resole.

Goodyear Welt Construction
(Drawing courtesy of Kastinger Berg-und Wanderschuhe)

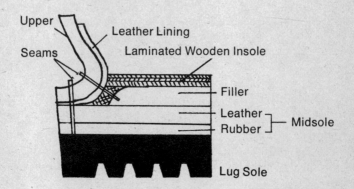

HOW TO MAKE SURE THE BOOTS FIT

After you have narrowed the search (on the basis of use) and have tentatively selected a model (on the basis of construction, weight, and price), there is one final consideration—fit. Every boot is constructed around a wooden or plastic form called a "last" which is designed to represent the average foot. The problem, of course, is that no one's foot is average. The majority of boots are made in Europe, and the average European foot tends to be narrower in the toe and broader in the heel than the average American foot. So European boots, in order to fit, must be made on American lasts. Because of the variation among lasts and sizes, it is sometimes difficult to get a good fit in the boot model you select. When this happens, it is essential to switch models. It is infinitely better to have a comfortable fit in your second-choice boot than a poor fit in what seems the ideal model. (The importance of a good fit means buying boots by mail is somewhat of a gamble.)

Most good boot salesmen will help you determine the quality of the fit, but there are five tests worth remembering: (1) As a general rule, if a boot is large enough it should be possible to place the forefinger behind the heel with the foot all the way forward and the boot unlaced. (2) With the boot snugly laced, your heel should be held down within the cup formed by the "heel counters" so that heel lift is no more than an eighth of an inch in normal walking. (3) Your toes should be free to wiggle, but neither the ball of the foot nor the heel should slip when the salesman anchors the boot to the floor and you vigorously try to twist your foot. (4) Boots should be snug in width but generous in length. (5) By kicking a post or hammering the floor with the toe, you can assure yourself that, even under load with swollen feet, your toes will never hit the end of the boot. If the boot passes these five tests and feels reasonably comfortable (allowing for newness), the fit is probably satisfactory.

Before the boots you purchased are ready for serious walking, they need to be treated and broken in. In comparatively dry, warm climates, applications of silicone compounds will improve water repellence and still allow the leather to breathe. In damp climates or in winter or spring, boots should be impregnated with a *silicone wax*, applied under heat, to prevent wetting. Oils are convenient for touching up seams just before and during trips, but should be generally avoided because of their tendency to overly soften and stretch leather.

Breaking in boots is a highly individual process. If feet vary in their resemblance to the average, they vary even more in their sensitivity to new boots. Some people go through agony breaking in supple trail shoes. Others can take stiff, new boots on a long trip without any discomfort after only a day or two of breaking in. The safest approach is to allow a month between purchase and the first serious walk—but start the breaking-in process right away.

Walking in the city is a good start. Hiking in the hills is much better. Running effectively hastens the process. So does bending and flexing and kneading new boots—especially if they bind. Fitting problems may need to be corrected by inserting arch supports or insoles, or by stretching the leather at points where the boot binds or pinches.

Some hikers make a regular practice of soaking new boots in water for five minutes, then hiking in them until they are dry to accelerate the process of conforming the boots to the feet. As a last resort, boots that stubbornly refuse to loosen up can be treated with oil. During and after the breaking-in period, appropriate heavy hiking socks should be worn. Thin socks won't provide protection and comfort. Too much sock (more than one heavy and one light pair) promotes friction inside the boot, which produces blisters.

Toughening up soft city feet can both hasten and minimize the breaking-in process. Walking barefoot toughens your feet. Concrete sidewalks effectively build callous, and sunlight and fresh air promote healthy feet. Some people harden skin by applications of alcohol. Athletes often toughen the soles of their feet by painting them with tincture of benzoate and dusting them with talcum powder. Toenails should be kept square cut and short, and never should be cut immediately before a trip.

When the right boot for the trip has been properly fitted, treated against the elements, well broken-in, matched to the right sock and snugly laced, you will be well prepared for the joys of comfortable hiking.

SOME GOOD VALUES FOR THE MONEY/BOOTS

Raichle Scout:

A moderate-weight, all-purpose trail shoe, the Raichle Scout is sturdy enough for heavy-duty backpacking, and light enough for trail hiking. The boot is made with split leather uppers and a yellow-spot Vibram sole. The Scout has a heavy-duty half-steel shank, and arch reinforcement. A very good sturdy and rugged shoe, with a suggested retail price of around $60.

Dunham Tyrolean Waffle Stomper, Models 6034 and 6035

This fine boot, which comes in both men's and women's sizes, is a very good all-leather lightweight trail shoe. It comes in suede, in narrow and medium widths, and has a Vibram sole and rolled scree guard. These models are always available in the Dunham Factory Outlet Store, as well as at most retailers. It usually sells for about $40 (less at the Factory Outlet).

Edge Boots

This is an excellent line of machine-crafted boots. Unfortunately, they're hard to find in the United States, but they're worth the search. The craftsmanship is among the best of any commercially

The New Trend in Lightweight Hiking Boots
(Photographs courtesy of L. L. Bean, Inc.)

Pathfinder, by American Footwear
(Photo courtesy of American Footwear)

made boots we've seen. If you travel through Canada, you'll probably be able to find them more easily. They retail for from around $70 up.

American Footwear Explorer

Full-grained leather, with padded collar and tongue, the Explorer also features a Montagnabloc Vibram sole. It is an excellent medium-weight hiking boot, easy to break in, that comes in men's and women's sizes. You should be able to get a pair for less than $50.

Danner Mountain Trail Boot, Model 6490

This very lightweight shoe has full-grain leather uppers, foam-padded ankle and tongue, a leather insole, steel shank, and a Vibram

Montagna lug sole. Stitched on a wide last it is an excellent boot for people with wide feet. It's one of the best-made hiking trail shoes, and has a manufacturer's suggested retail price of just under $70.

Summit High Noon

A split-leather lightweight hiking boot with padded uppers and tongue, this boot has a steel shank, reinforced toe, and mid-sole. It comes in men's and women's sizes, but only in a medium width. A fine boot, under $50.

Summit Tenaya

A rough-out, medium-weight hiking boot with a Vibram Security sole, the Tenaya has double-leather mid-soles, full steel shank, and reinforced toe and heel counter. Padded uppers and tongue make

Raichle Heavyweight Hiking
and Climbing Boot
(Photo courtesy of Raichle)

Summit High Noon
(Photo courtesy of
Kalmar Trading Corporation)

Summit Tenaya
(Photo courtesy of
Kalmar Trading Corporation)

the boot easy to break in. This shoe comes in men's and women's sizes too. A very durable hiking boot that costs well under $50.

Galibier Vercors

Galibier makes a very easy-to-fit hiking boot that runs wide. The Vercors has a partial steel shank that ends at the instep. It features reverse tan two-piece uppers with watertight stitching and a Jannu Galibier sole. This is the most expensive boot that we've listed, but it's of very high quality. If you find this on sale, buy it fast.

HOW TO WINTERIZE YOUR BACKPACKING BOOTS

If you are planning to buy boots that are going to be used all year round, particular attention has to be paid to the fit: The boot must not be too tight. This is *very* important, as tight-fitting boots during winter backpacking trips can invite frostbite. If you intend to use your boots during the winter, first try on the boots in the store, wearing two pairs of socks. One pair of socks should be a heavy-duty Norwegian Ragg type, and the other can be a medium-weight wool sock.

You'll still need more on your feet during the winter than just two pairs of socks and your boots. Supergaiters, or an overboot, are essential for winter hiking. We've never seen Chouinard Supergaiters on sale, but you *can* get overboots for less than manufacturer's list price. Some good overboots are: the Eddie Bauer Downfill Overboot, the Rivendell Overboot, and the REI Expedition Overboot.

You can also achieve additional insulation by adding ¼-inch ensolite between the Vibram sole and the overboot. To make your own ensolite foot warmers, just trace the outline of your boot on a small sheet of ¼-inch ensolite and cut out.

If you're going to be using crampons and all you have are flexible-soled boots, you might try this trick from Chris Rowins, one of our climbing partners: Trace the outline of your boots on ¼-inch plywood. Using a jigsaw, cut out the two pieces of wood. Place each piece of wood underneath the boot before attaching your crampons. Chris has climbed at a very high standard using this money-saving device.

HOW TO ADD VIBRAM SOLES TO FLAT-SOLED BOOTS

If you have a favorite pair of gum-soled boots but they don't give you the traction you need on steep trails, try adding a Vibram sole to them. We suggest that you take your boots to your local shoe repair person and ask if he or she can replace the sole with Vibram. Many cobblers have the facilities to do this, and it's not a complicated or expensive process. If your local cobbler won't or can't do it, then we recommend that you send them to one of the specialized backpacking and climbing cobblers listed later in this chapter.

HOW TO WATERPROOF YOUR BOOTS

Good hiking boots will give you many years of service if you take care of them. One essential to good care is waterproofing them.

The two best waterproofing compounds are Sno-Seal and Tecnica Waterproofing. Both are wax-based, containing no oils and little or no silicone, both of which inevitably break down the leather or affect the glue. We never recommend the use of spray silicone waterproofing on any boot; it just doesn't help the life expectancy.

HOW TO CONVERT MOUNTAINEERING BOOTS INTO CROSS-COUNTRY SKI BOOTS

You can convert your mountaineering boots into cross-country ski boots by purchasing a mountaineering binding. The two bindings that are the least expensive and the easiest to attach are made by Selewa and by Tempo. Follow the very simple directions provided in the binding boxes and you'll be able to cross-country ski within fifteen minutes—with much less danger of getting frostbitten toes and without the expense of buying cross-country ski boots.

HOW TO MODIFY LACE-UP SKI BOOTS FOR BACKPACKING AND WINTER MOUNTAINEERING USE

Christopher Ellms, a North Country Mountaineering Guide, came up with this idea: If you have a pair of loose-fitting lace-up boots (or if you buy a pair cheaply in a local thrift shop), have a cobbler remove the bottom sole and replace it with a Vibram sole. The cobbler will have to trim the front of the ski boot so that it's circular, not square. Used with a Supergaiter or a good overboot, it makes an excellent stiff technical rock and ice climbing boot. Chris successfully used these boots for two seasons of New England ice climbing, with never once having cold feet. The cost of the boots and resoling was $14.

Habeler Superlight Mountaineering Boot (Photo courtesy of Kastinger Berg-und Wanderschuhe)

Royal Robbins Alpine Boot (Photo courtesy of Robbins Mountain Shop)

Dunham Mountain Boots
(Photo courtesy of Dunham Shoe Factory)

RECONDITIONING OLD HIKING AND BACKPACKING BOOTS

A backpacker's best friend is his or her boots. When your friend is sick, worn down, needs to have its seams rebound, or is in need of a fresh sole, the services of a professional boot surgeon are needed. This is one instance where it pays to spend a bit of money —and to avoid your local shoe repair shop. The following cobblers have had a great deal of experience helping to mend sick and wounded boots. Always write for an estimate first.

DIRECTORY OF COBBLERS EXPERIENCED IN REPAIRING BACKPACKING AND MOUNTAINEERING BOOTS

All of the cobblers on the pages that follow have been highly recommended to us by fellow backpackers. The cobblers with asterisks next to their names are people with whom we've personally done business and whose work we can recommend from our own experience.

Alaska

Two Wheel Taxi and Ski Shop
405 East Northern Lights Boulevard
Anchorage, Alaska 99504
907-278-4578

Arkansas

Ozark Mountain Sports, Inc.
226 North School
Fayetteville, Arkansas 72701
501-521-5820

California

Alpine Outfitters, Inc.
1538 Market Street—The Mall
Redding, California 96001
916-243-7333

Goodwin—Cole Sports
1315 Alhambra Boulevard
Sacramento, California 95816
916-452-6641

Mountain Traders
1711 Grove Street
Berkeley, California 94709
415-845-8600

Ritz Shoe Rebuilders
1869 Solano Avenue
Berkeley, California 94907
415-526-3316

Robbins Mountain Shop
1508 10th Street
Modesto, California 95350
209-529-6917

Sierra Outfitters
4915 Fulton Avenue
Sacramento, California 95821
916-481-2480

Tetons West Mountain Shop
87 East Blithedale Avenue
Mill Valley, California 94941
415-383-4050

Tetons West Mountain Shop
943 Sir Francis Drake
Kentfield, California 94904
415-457-8780

Colorado

The Cobbler
1702½ West Colorado Avenue
Colorado Springs, Colorado 80206
303-475-7626

Colorado Shoe Company
3103 East Colfax Avenue
Denver, Colorado 80206
303-355-1991

Leroy Gonzales
Telemark Ski and Mountain Sports
416 East 7th Avenue
Denver, Colorado 80205
303-837-1260

Steve Komito*
Davis Hill
Box 2106
Estes Park, Colorado 80517
303-586-5791

Neptune Mountaineering*
1750 30th
Boulder, Colorado 80301
303-442-3551

Table Mesa Shoe Repair
665 South Broadway
Boulder, Colorado 80303
303-494-5401

Georgia

Rocky Mountain Sports
6125 Roswell Road
Atlanta, Georgia 30328
404-252-3157

Idaho

Mountain Boot Repair*
P.O. Box 94
Ketchum, Idaho 83340
208-726-9935

Massachusetts

Strand's Ski Shop, Inc.
1 West Boylston Drive
Worcester, Massachusetts 01606
617-852-4333

Michigan

Bill and Paul's Sporthaus
3645 28th Street S.E.—Eastbrook
Mall
Grand Rapids, Michigan 49508
616-949-0190

Montana

Schnee's Boot Works*
411 West Mendenhall
Bozeman, Montana 59715
406-587-0981

Missouri

Ride-On Outdoor Sports
3959 Broadway
Kansas City, Missouri 64111
816-753-2900

New Jersey

Hills n Trails
93 Brant Avenue
Clark, New Jersey 07066
201-381-9046

New Mexico

Mountains and Rivers
2320 Central S.E.
Albuquerque, New Mexico 87106
505-268-4876

New York

Boot Repair
32 Bedford Park Boulevard
Bronx, New York 10458
914-776-1559

Down East*
93 Spring Street
New York, New York 10012
212-925-2633

Ohio

Rube Adler Sporting Goods
11642 Detroit Avenue
Cleveland, Ohio 44102
216-226-1740

Pennsylvania

The Mountain Trail Shop
5435 Walnut Street
Pittsburgh, Pennsylvania 15232
412-687-1700

Utah

Starlight and Storm—Mountain Boot
Repair
3288 South 13th East
Salt Lake City, Utah 84106
801-466-6714

Washington

Dave Page Cobbler
346 Northeast 56 Street
Seattle, Washington 98105
206-523-8020

3

HOW TO BUY A NEW TENT

─────────────────────────────────── **About Tents**

Backpacking tents used to be of very simple design. A cotton-canvas tarpaulin was folded in half and suspended between two vertical poles. This produced the simplest and least expensive A-frame mountain shelter.

Today tentage has undergone radical changes and improvements. Very lightweight ripstop nylon and Gore-Tex® cloth* have been combined with machined aluminum tubing to produce superior tent shelters.

The most prevalent shapes of today's backpacking tents are: the A-frame, dome tents, external frame (Blanchard-type) tents, and tepee or pyramid-shaped tents.

Most outdoor stores will have examples of these types set up on the selling floor. Take your shoes off, get into the tent, and see what it's like from the inside out. It's most important that you like the shape (and color) of the tent you buy because it is one of the most expensive items you'll buy, and you'll probably use it for many years to come.

If the tent in which you're interested isn't set up, ask the store manager or clerk to set it up for you. He or she will rarely object. If for some reason he refuses, be sure he agrees that it's all right for you to return the tent if you should discover some defect when you get it home and set it up there.

ADVANTAGES OF THE A-FRAME TENT

The A-frame is one of the simplest tents to set up. It is lightweight, easy to pack, easy to modify and repair. Some A-frame tents come with features like a winter crawl passage, a frost liner, a cookhole, and shock-corded poles.

A winter crawl passage is a sleeve of nylon through which one gains access to the tent without bringing snow inside. People usu-

*See Chapter 11 for information about Gore-Tex®.

ally take their boots off here in order not to get the inside of the tent wet. A cookhole is simply an opening on the floor of the tent where you can set up your stove without damaging the tent floor. Shock-corded poles are aluminum poles strung with elastic cord which keeps them together to prevent loss. The shock-cording also makes it much easier and faster to set up the tent, especially during severe weather.

ADVANTAGES OF DOME TENTS

Because its pole system fits into self-contained sleeves that require no guylines to set up to support the tent, you can move a dome tent without having to dismantle it. This is particularly handy if, as sometimes happens, a tent has been set up in a water channel.

Because of its circular shape, the usable inside space of a dome tent is greater than that of a conventional A-frame tent.

ADVANTAGES OF EXTERNAL FRAME TENTS

The original external frame tent was designed by Smoke Blanchard and used successfully during the 1963 American Everest Expedition. A number of manufacturers, including Gerry, Bishop's Ultimate Outdoor Equipment, and Eureka, have been using adaptations of the Blanchard design for years now. The external frame tent is as movable as a dome tent and is, because of its external skeleton, structurally very strong. The extra poles and other attachments do, however, add extra weight.

ADVANTAGES OF TEPEE AND PYRAMID-SHAPED TENTS

Tepee or pyramid-shaped tents provide extra headroom and more usable inside area than an A-frame tent. This shape of tent also tends to be very stable in snowy conditions—more stable, in fact, than the A-frame. Some tents of this design still rely on an internal single pole; these tents are of the Logan type sold by Alpine Designs and REI. Though these tents are inexpensive and roomy, an inside pole inevitably gets in the way and effectively cuts down usable floor space.

The Jansport Rover Dome Tent
(Photo courtesy of Jansport)

Bishop's Ultimate Tent with Fitted Fly
(Photo courtesy of Bishop's Ultimate Outdoor Equipment)

TENT

FITTED FLY

FOURTEEN QUESTIONS (AND ANSWERS) TO HELP YOU DECIDE WHAT KIND OF TENT YOU NEED

Everyone shopping for a new backpacking tent ought to ask himself or herself the following questions. These should help you decide what kind of tent you need and what you should look for before making this major equipment decision.

1. Do I need a tent for summer use only?

If you need a tent for summer use only, a simple, single-pole suspension tent will be perfect for you. Preferably it should be waterproof and have a mosquito net opening to allow adequate ventilation on warm summer evenings. The Eureka Catskill Tent, the Trailwise Wind River Tent, and the Nu-Lite 2-Man Tent are all relatively inexpensive tents that offer these features.

2. Do I need a tent for both summer and winter use?

A winter tent should have a waterproof floor and waterproof sidewalls extending 18 inches up the sides. It should have a crawlway entrance with a sleeve that allows you to take off your boots without actually entering the tent. The tent should have a very good pole support system which will not collapse under snow loads.

The least expensive winter tent with the above features will cost you about twice the price of a decent summer tent. But you can convert a summer tent into a winter tent if you are willing to do some modification. Some reasonably priced tents that can be used for both winter and summer are the Camp Ways Caribou Mountain Tent and the EMS Alcove Tent.

3. Do I need room for more than two people? Do I want a family tent?

You may not have a family, but you may like the feeling of a roomy tent. When we go backpacking, our choice is a three-person tent because it offers room enough to store all our equipment inside, out of the rain and snow, and enough space for us to spread out comfortably. If you're backpacking with two people, the difference in weight—for a whole lot of comfort—is less than one pound per person.

Two good economical choices would be: the 3-person Sierra Designs Pleasure Dome or the REI Nylon Camper.

4. Do I want a tent that's very easy to set up?

Generally, the easiest tent to set up is the A-frame. Remember, the bigger the tent, the harder it is to pitch, and the longer it takes.

Eave Tent
(Photo courtesy of Moss Tentworks, Maine)

5. Am I willing to make my own tent?

There are about a dozen tent kits on the market, some of which are actually quite easy to make. This is the most economical way to obtain a good tent at a good price. For more information on these kits, please see page 65, as well as Chapter 12.

If you are a really ambitious sewer, you can make your own tent from scratch. By the time this book is published, Recreational Equipment, Inc., Seattle, Washington, should have introduced their new tent patterns. They also sell ripstop nylon by the yard.

There are at least two recent books on the market that offer information on making your own tents: *Make Your Own Camping Equipment* by Robert Sumner (Drake Publishers, $6.95) and *Lightweight Camping Equipment and How to Make It* by Gerry Cunningham and Margaret Hansson (Scribners, $5.95).

6. Am I willing to modify a tent I buy?

If you're willing to make modifications in a store-bought tent, there's no end to how deluxe your tent can be. Later in this chapter we'll tell you how to install your own cookhole, how to add A-frame poles to single-pole tents, and how to add snowflaps.

7. Do I want a tent that has resale value?

The longer we go backpacking, the more attached to certain brands of equipment we get. We recommend that you purchase your tent with an eye toward the future. Eventually you're going to want to get a better tent. If you buy a medium-quality tent now, and maintain it, you can probably resell it for almost what you paid for it, given the current rate of inflation.

8. Do I want an A-frame tent, a dome tent, or a teepee tent?

This is a matter of personal taste. Any tent that functions in a given situation is the correct tent. For more information on these types of tents, please see pages 49 and 50.

9. Do I need a rain fly?

If the fabric of the tent is not waterproof, you *must* get a tent with a rain fly. Most tents today come with this feature, with the exception of the new Gore-Tex® tents (see Chapter 11). Rain flies, purchased separately, are very expensive. If your tent does not have this important protection, a large sheet of plastic over the tent will work effectively for a short period of time, but you must expect heavy condensation, since plastic doesn't breathe.

10. Do I need a tent of waterproof fabric?

Some tents are completely waterproof. These are usually among

The Rivendell Gore-Tex® Bombshelter Tent
(Photo courtesy of Rivendell Mountain Works)

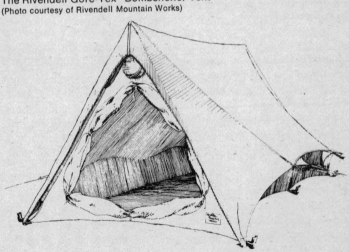

the most inexpensive to purchase (with the exception, of course, of the new Gore-Tex® tents). With adequate ventilation, these tents are great. Tents that do not feature waterproof fabrics require a rain fly or a plastic sheet over them.

11. Do I need a waterproof floor?

We strongly recommend a waterproof floor. Otherwise you'll wake to find the dew has come up through the floor of your tent and dampened your sleeping bag and all your belongings.

12. Do I want mosquito netting?

In buggy areas, mosquito netting is a blessing. It also allows good ventilation because you can sleep with your door open. If you know you are going to be backpacking where bugs are not a nuisance, a waterproof tarp may be all the tentage you need.

13. Do I want a cookhole?

You need a cookhole only if you intend to go winter mountaineering.

14. Do I want shock-corded poles?

Only the most inexpensive tents come without shock-corded poles. Shock-cording does slightly increase the time needed to set up a tent, and it increases the carrying weight. Many experienced backpackers deliberately remove the shock-cording from their tent poles for summer use.

SOME GOOD VALUES FOR THE MONEY/TENTS

EMS Alcove Tent

Designed for winter mountaineering, this tent has a zipper entrance on one side and a crawlway alcove entrance on the other. It comes with a cookhole, rain fly, poles, pegs, and stuff sack, and accommodates two people. A very good buy for about $135, this is a multiseason A-frame tent that is good for all weather conditions.

Eureka Catskill Tent

This two- or three-person tent has an A-frame pole in the front and a single rear pole, in addition to a zippered door and a zippered window that provide good ventilation. Another plus is a rain fly. The Eureka tent comfortably sleeps two people and sells for less than $70.

Trailwise Wind River Tent

Made of waterproof-coated nylon (no rain fly is necessary). Single-pole pup, designed to sleep two people, this tent uses single poles on both sides. It is very light (3 pounds, 2 ounces) and costs under $60.

REI Economy Nylon Tent

This is one of the very best beginner backpacker tents for the money. It is a single-pole tent—very easy to set up—that can be purchased for less than $35. It sleeps two people. It is made of lightweight water-repellent nylon, with a mosquito net door.

Nu-Lite 2-Man Tent

Made by the Eureka Tent and Awning Company, this shelter is made of waterproof-coated nylon taffeta. It has a zippered door covered with mosquito netting, and single-pole suspension. It weighs 4 pounds, 2 ounces and is priced at under $40. A perfectly serviceable tent.

Kalmar Trading Corporation's Sitka Tent

Kalmar makes a very roomy, ruggedly built, waterproof tent that should last many seasons. It has single-pole suspension, with side-pole pullouts, and comes in two- and three-person sizes. Other features are a zipper entrance door, and rear window. Prices vary from store to store, but it's usually priced under $100.

Coleman Model 838

A two-person tent recently introduced by Coleman, Model 838 is an extremely roomy A-frame tent with a horizontal bar that connects the front and rear A-frame. This is a very stable tent. Comes with a full waterproof fly. You ought to be able to get it on sale for about $90.

World Famous #486 2-Person Nylon Tent

This tent, made of waterproof nylon, has single-pole suspension and a mosquito net door, among other features. It retails for around $40. World Famous produces a very good line of summer backpacking tents that sell for under $90. It's not a "name" brand—but it is a *good* brand.

Camp Ways Caribou Mountain Tent

A fire-retardant two-person tent that comes with a rain fly, the Caribou has 15-inch coated sidewalls, a tunnel entrance at one end, and a zipper entrance at the other (for very good ventilation). It weighs around 7 pounds and sells for around $115. A reasonably priced winter mountaineering tent.

WHAT TO LOOK FOR IN A TENT FOR WINTER BACKPACKING
Waterproof Sidewalls

All winter tents should have waterproof sidewalls extending at least 12 to 18 inches on all sides; these are absolutely essential in a winter tent. Their purpose is very simple. When you set up your

Coleman Tent Model #838
(Photo courtesy of the Coleman Company)

winter tent, chances are it'll be on snow you've flattened down with your snowshoes. But it's extremely likely that more snow will fall while the tent is set up. This snow will accumulate in drifts against the sides of the tent. Once you are inside the tent, of course, the air inside will warm up, and the snowdrifts against the tent walls will melt. If the sidewalls are not waterproof, this melting snow will come right inside your tent.

Crawlway Entrance or Sleeve

This feature allows easy access into your tent without bringing a blizzard inside. It also provides an area for removal of snow-covered boots. A longer crawlway will also provide extra equipment storage space, and will consequently make your sleeping area more comfortable.

Frost liner

During winter, condensation freezes on the inside of your tent. A frost liner will keep condensation from dripping on you and your equipment.

Snow Flaps

Snow flaps are rectangular pieces of nylon material sewn to the bottom edges of your winter tent. They help anchor your tent in windy conditions, making it much more stable.

Shock-Corded Guylines

Shock-corded poles are a *necessity* in a winter tent. They simplify erection of the tent in a cold, hostile environment. They also prevent loss of pole sections.

Cookhole

During prolonged forced stays inside your tent due to harsh winter weather, you'll definitely want a cookhole. It is the only *safe* way to cook without melting your tent fabric. For details on installing a cookhole yourself, see page 59.

Ventilation Hole

Because moisture and noxious gases produced by camping stoves accumulate inside the tent, you *must* have a ventilation hole in your winter tent. If the tent you find at a bargain price doesn't have a ventilation hole, you can make one yourself. The books mentioned on page 53 contain instructions on this simple procedure. Some equipment modification specialists like Leon Greenman at Down East, 93 Spring Street, New York, New York, will install a ventilation hole for you for under $10.00.

Equipment Pockets

Most high-quality winter tents—for example, those manufactured by Sierra Designs, North Face, and Trailwise—come with equipment pockets. But these are very easy to install. Cut out a pocket the size you want from mosquito netting or ripstop nylon and attach it along the seamed joint where your waterproof sidewall ends and the tent wall begins. Equipment pockets are particularly important if you wear glasses (and everyone out during winter conditions should be using snow goggles to prevent snow blindness); the pockets should prevent accidental loss of the eyeglasses, or accidental breakage.

Snow Pegs

These usually are purchased separately. You can buy them in most backpacking retail shops for about 90¢ each. They're well worth the money because they provide the best anchorage system for your tent during snowy conditions.

Seams Sealed Inside and Out

Most tents will need to have their seams sealed. K-Kote seam sealant will do the job if you follow the directions on the tube; then you won't have to worry about wind, rain, or snow penetrating your tent.

(Photo courtesy of Kenyon Industries)

HOW TO MODIFY A SUMMER TENT INTO A WINTER TENT
How to Install Your Own Cookhole:

Basically, a cookhole in a winterized tent is simply an aperture in the floor of the tent where the stove rests while in use. This prevents damage to the tent fabric from sparks, heat, food, or spilled fuel. But once the stove has been put away and the fire is out, cold or moisture can make life miserable unless the tent floor is made whole again. The solution: a hinged flap which can be resealed again. Here's how to do it:

MATERIALS NEEDED

ruler

sharp knife

scissors

felt-tip pen

1 tube K-Kote seam sealer

1⅓ yards 1-inch Velcro

16″ by 16″ square of ripstop nylon

Pitch your tent inside your house or outside in your yard. Measure back 12 inches along the floor from the front of your tent and with a felt-tip pen make a mark there. Measure the width of your tent and mark the midpoint to correspond with your 12-inch mark. Now, using a ruler, mark off a square 12 by 12 inches with the midpoint of one side of the square corresponding with the midpoint mark you've already made. What you should have now is a drawing of a square that is nicely aligned with the entrance to the tent, and twelve inches back from it.

Inside the square you have measured, measure off another

square that measures 10 by 10 inches, with a two-inch border on each side.

Using a sharp, heated knife, cut the *inside* square. Reheat the knife as necessary.

What you should have now is a 10- by 10-inch hole with a 2-inch border on each side.

Now, at each corner, using your hot knife, make a cut 2 inches deep that dissects each corner at a 45-degree angle. Fold these flaps under twice, and sew in place on a sewing machine. You'll wind up with a seam 1 inch wide all around your new cookhole.

Cut 1-inch Velcro tape into strips that fit three sides of the perimeter of the hole. Sew into place. The side nearest the tent entrance should have Velcro sewn on it.

Out of a separate piece of nylon, cut a square 16 by 16 inches. Fold each side over 1 inch and sew down. Sew Velcro to three sides of the square.

Place the larger square onto the cookhole so that the Velcro borders match up.

Sew the fourth side (the side furthest away from the tent entrance) down.

Now you can open and close your cookhole with ease, and it's almost ready for use.

To get it ready for use, turn the tent over and use K-Kote to seal all the seam stitching.

How to Add Snow Flaps to an A-Frame Tent

Measure the length of your tent. If there is a seam along the edges of the tent, very carefully open the seams 6 inches from each corner, taking care not to cut or damage the fabric. Sew into the seam a strip of ripstop nylon 10 inches wide by 3 feet long (the edges of which have been sealed by a hot knife). Do this to both sides, adding two snow flaps to each side of the tent.

You can also add snow flaps to the front of the tent, but these snow flaps should be one long piece and custom-measured to the tent.

Please see the illustration on page 61.

How to Add a Crawlway or Winter Vestibule

This is the hardest tent modification for an inexpert sewer. Mistakes can be costly. Since there is currently no kit available for a winter vestibule, and because even how-to-make-your-own-equipment books don't go into this touchy subject, we recommend that you have an expert do this for you. The nicest modifications of this type that we've seen have been done by Down East, 93 Spring

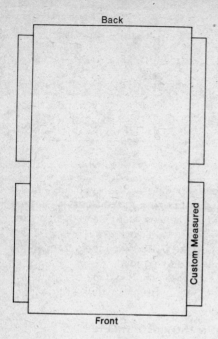

Back

Custom Measured

Front

Street, New York, New York. You can expect to wait a couple of weeks to have it done, and to pay about $40 to $45. The Alpine Guild, at 300 Queen Anne Avenue North, North Seattle, Washington 98109, will also do this alteration for you. Write to them for an estimate before you send your tent.

HOW TO ADD A-FRAME POLES TO SINGLE-POLE TENTS

This modification will give you easier access into the tent, and give the shelter more stability. It'll also increase its resale value.

Measure each side of the tent. Then add two inches to each total. You will need twenty inches of 1-inch nylon webbing, four 4-inch lengths and one 6-inch length; a set of A-frame poles (or individual aluminum pole sections that will equal the distance you already measured), and an A-frame coupling joint (about $2.00).

Two-thirds of the way up each side of the tent open the seam about an inch and a half. Fold the 4-inch webbing in half, placing

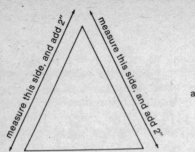

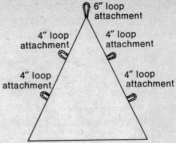

the ends into the 1½-inch seam opening, and sew in place. Repeat the same procedure one-third of the way up the sides of the tent. You now have two loops sewn in place on each side of the tent. See the illustration above.

What you'll do now is set up the poles, place them through the loops, and join them, using the coupling joint.

Now take the 6-inch piece of webbing, fold it in half, and sew it to the crown (tip) of the tent. Seal the seams with K-Kote.

HOW TO CONSTRUCT AN IMPROVISED SHELTER

If you are a first-season backpacker, don't rush out to spend $35 to $200 for a tent. You are best off borrowing or renting one, but if you can't, or if you just want the independence of having your own shelter, we recommend that you try the following:

Tarp Tents

Purchase a Griffolyn reinforced tarp, 8 feet by 12 feet. This tarp is made of plastic sheeting reinforced with crossed nylon thread and is very strong. It comes with 20 Versa Ties, which will grab any part of the tarp, allowing you to pitch the tarp in many different ways for use as a rain shelter. The package illustrates the various ways to pitch the tarp. It makes an economical summer shelter for under $15.

Another good tarp is the EMS Tarp, which sells for under $30. This tarp is constructed of 2.2-ounce ripstop nylon. Three grommets are attached along the 9-foot side, and 5 grommets are attached along the 10-foot side for flexibility in rigging.

The REI One-Man Instant Tent is a 9-foot tube with a 3-foot 3-inch diameter. It's made of polyethylene and is completely waterproof. It can be strung up between two trees for a very inexpensive (under $7), practical one-person shelter.

The Lashed Lean-To/A Warning

When we were Scouts and at summer camps, the most popular summer shelter was the lashed lean-to with a covering of boughs. We do not recommend this type of shelter today; the lashed lean-to requires either a dense selection of deadfall, or freshly cut young trees. Our woodlands today don't need any further damage. Carry a tarp and protect the natural environment.

HOW TO PROLONG THE LIFE OF YOUR TENT

The life expectancy of a tent is directly related to the care you give it. Even a very inexpensive tent will last a long time if used properly, and cared for.

There are a few simple ways to make any tent last longer. Be sure that after each use the bottom of the tent is brushed and sponged off with lukewarm water. If the tent is moist or wet, turn it inside out and hang it to dry. Make sure the seams are properly waterproofed with K-Kote seam sealant. Take care not to pitch the tent on any sharp or projecting rocks or roots. Don't overtighten guy lines; this causes unnecessary stress on the seams. Check all zippers and mosquito netting upon returning home and repair any rips immediately. Carry your tent pegs in a separate nylon stuff bag so that they don't tear or muddy the tent. Roll all poles in the inside of the tent before placing it in the stuff sack; this will protect your surprisingly expensive poles from damage. If a tear develops in the tent, it should be patched in the field with ripstop repair tape; later, when you return to home base, you can sew the tape down for a more permanent repair.

The Little Amenities/A Sponge

We also suggest that you take a sponge with you when you go winter backpacking and camping. This will help clean up inadvertent messes before they dry and become difficult problems.

The Little Amenities/Space Blanket

In addition, we recommend that you buy a space blanket for winter backpacks. Before you pitch your tent on a snow field, put the space blanket reflecter side up onto the snow. It will add extra insulation and keep the bottom of your tent dry. Using the space blanket (or even plain plastic sheeting) in summer will keep dirt from being ground into the bottom of your tent.

We would recommend that you *never* cook inside your tent: The danger of fire is too great. This past fall we witnessed the total destruction of a St. Elias tent on Mount Washington. Incidents like this have become too frequent. A winter mountaineer may of necessity have to cook inside his or her tent—but there is no good reason why a three-season backpacker should do the same.

HOW TO DO MINOR TENT REPAIRS

Burns, Rips, and Tears

You can sometimes find a new tent at a clearance center or bargain basement because it has a minor rip or tear (we define a minor rip as a clean tear up to 6 or 7 inches long). You should have no problem getting a discount on a new, slightly ripped tent. Or you may find an otherwise good used tent with a small cigarette hole burnt in it.

Don't worry about these minor flaws . . . Ripstop nylon repair tape with adhesive backing, available at most backpacking equipment stores, will take care of the problem. The tape costs about $1.00 or $1.50 per roll, and comes in several colors.

For best results, sew the tape to the tent and use K-Kote seam sealer over the stitching line.

How to Replace Burnt or Torn Nylon Panels

Very carefully open all the seams connected to the panel. Carefully remove the panel in one piece and place it on a piece of ripstop nylon replacement fabric. Trace the outline of the damaged panel onto the new fabric, cut and sew in place. Seal seams with K-Kote.

How to Replace Broken Zippers

Carefully remove the whole zipper. Bring it to your local backpacking supply store, so you get a proper replacement, and sew in place as you would a regular zipper on a garment. If you are a tentative sewer, or are afraid of zippers (as some sewers are), ask your local backpacking retailer if there is someone in your area who does minor repairs, or consult the who-does-repairs sections in Chapter 10. You may want to shop around in order to compare prices.

How to Replace Torn Mosquito Netting

Cut away the area that is torn. Buy a mosquito netting replacement patch, and sew it on by hand.

How to Rewaterproof Your Tent

Purchase a quart of nylon waterproofing solution. Carefully clean the inside surface of the tent, and apply the waterproofing solution as directed, in a very well-ventilated room. You ought to waterproof your tent every season.

Sealants

We recommend that each year you go over all tent seams with K-Kote tent sealant.

TENT KITS AND WHERE TO BUY THEM

Making your own tent requires patience and some sewing skill. You don't have to be an expert sewer though to make a fine tent from a kit. We know one young man, age 15, a student at the White Mountain School in Littleton, New Hampshire, who had never sewn before but bought himself a Frostline Tent Kit—and wound up with a very serviceable tent. If you make your own tent from a kit, you can expect to save about 40% of the usual cost.

Most of the kit producers listed below have wide distribution of their products. If you can't find their kits where you live, write to them directly.

Altra, Inc.
3645 Pearl Street
Boulder, Colorado 80301

Eastern Mountain Sports Kits
1041 Commonwealth Avenue
Boston, Massachusetts 02215

Frostline Kits
452 Burbank
Broomfield, Colorado 80020

Holubar Outdoor Equipment Kits
Box 7
Boulder, Colorado 80306

Mountain Adventure Kits
P.O. Box 571
Whittier, California 90608
213-698-7311

Plain Brown Wrapper, Inc.
2055 West Amherst Avenue
Englewood, Colorado 80110

4

HOW TO BUY A NEW SLEEPING BAG

———————————————— **About Sleeping Bags**

BACKGROUND INFORMATION/RECENT RISES IN THE PRICES OF NYLON AND DOWN

Sleeping bags have always been a major investment for the camper and backpacker, but within the past three years prices have risen dramatically. The price of down was the first to increase—by as much as 20 percent. Then, during the winter of 1977, the cost of all nylon rose by 10 percent. That spring, down prices went up an additional 25 percent. Altogether, inflation has imposed a 55 percent increase in the cost of a down sleeping bag over less than three years! Lucky are those among us who have down sleeping bags turning into gold in our closets. But for those of us in need of new sleeping bags, this is the time when careful buying has to be the *only* way to buy. The first step is to learn the basics about sleeping bags and determine what kind of bag you need. You may decide that you don't require a down-filled sleeping bag, that a bag filled with Hollofil II or Polarguard will serve your purpose just as well. You will also have to decide what shape bag you prefer.

There are three basic shapes of sleeping bags: rectangular, modified mummy, mummy.

ADVANTAGES OF THE RECTANGULAR SLEEPING BAG

From our point of view, one of the major advantages of this bag is that the manufacturing cost is low, and consequently so is the retail price. Rectangular sleeping bags are available at most discount department stores, hardware stores, military surplus stores, and Boy Scout supply depots. They're usually made with Dacron polyester fill, and they generally cost from $12 to $20. Steve used a bag like this through his first six years of Scouting, and it served admirably during spring and summer use. The only disadvantage was its weight—about 7½ pounds. Dacron-filled rectangular bags weigh more than other sleeping bags on the market.

Some manufacturers produce Polarguard-filled and/or Hollofil-stuffed sleeping bags in this shape, primarily for people who dislike mummy-shaped bags. These bags are much lighter (because the Dacron filament is actually hollow) than the old Dacron bags, but not quite as light as those filled with down.

The only real advantage of the rectangular bag is the foot room it offers people who don't want to confine their feet.

ADVANTAGES OF THE MODIFIED MUMMY SLEEPING BAG

The modified mummy is more tapered than the rectangular sleeping bag. As a consequence, there's less dead air space for the body to heat. The modified mummy is also lighter to carry than the rectangular bag.

ADVANTAGES OF THE MUMMY SLEEPING BAG

This is the most efficient shape bag for a backpacker. It's the lightest possible bag, and it has the most efficient heat retaining system. The bag completely covers the body, including the head, preventing heat loss. The mummy bag has become the standard in the backpacking world.

In the past almost all mummy bags had a zipper down the front, but today it's common for bags of this shape to have zippers down either side. This allows two mummy bags to be zipped together.

THE SYNTHETIC FILLS/DACRON 88

The insulating material in these bags is the heaviest of all synthetic fills. Nevertheless, it produces very good insulating characteristics and is commonly used in the less expensive sleeping bags. Dacron 88 bags are perfect if you're car camping, or if you don't mind backpacking the extra weight.

THE SYNTHETIC FILLS/DU PONT HOLLOFIL II ™

Hollofil II used to be called Fiberfill II. Whichever name is used, the material is the same: a hollow, short-stranded filament of man-made Dacron polyester fiber. This fiber has lofting characteristics similar to down: The material can be compacted into stuff sacks; it regains its loft easily; and it's so dense that it creates an efficient dead air space around the sleeper. Because the fibers are short stranded, and hollow, the weight of the insulation as compared to bags marked simply Dacron polyester is 17 percent less per square inch of insulation. The Hollofil fiber has a long life expectancy when properly maintained.

Hollofil II is superior to down filling under rainy or wet conditions. Down will become matted when wet, and lose virtually all of its insulating properties. On the other hand, a wet sleeping bag made with Hollofil II can be wrung out and used; the nylon covering fabric will still be damp, but the insulation itself will not retain substantial moisture. Natural body heat will dry the nylon as you rest.

HOW TO CLEAN AND MAINTAIN HOLLOFIL II SLEEPING BAGS

Items meeting labeling, fabric, and interlining specifications for, and filled with, Dacron polyester Hollofil II should be laundered using a mild soap or detergent. Do not dry clean.

Here are DuPont's instructions for cleaning and caring for Hollofil II sleeping bags.

How to Hand Wash a Hollofil II Sleeping Bag

Soak in lukewarm water with a mild soap or detergent, pressing by hand until fully submerged. Continue to press or flush water through until clean. (If item is unusually soiled, it may require a second washing.) Rinse in clear water removing the soap and soil. Do not wring the garment or bag; press the water out by hand.

Air dry by hanging over a line. Dacron absorbs less than 1.0% moisture, and the fiber will dry quickly. The moisture will run to the bottom of the bag or garment after hanging for a short while. Drying can be speeded by pressing this water from the bottom by hand.

How to Machine Wash and Dry Hollofil II Sleeping Bags

Machine wash in tumble washers only. The machine should be the heavy-duty, front-loading type normally found in laundromats. Use mild soap or detergent, and warm, gentle cycle. Agitator and/or home washing machines should not be used, since damage to the bag, garment, or machine could result.

These can be dried in the same manner as with hand washing (over a line), or machine tumble-dried. The dryer should be the large, heavy-duty type normally found in laundromats. *Use low heat*. More than one cycle may be required for complete drying.

Extra care should be taken when machine washing and drying. Damage can be minimized if the washing machine is fully loaded, because tumbling is reduced. Wash and dry items with zippers closed. Safety-pin the filling and outer fabrics together if the outer fabric is not quilted or sewn to the filling.

How to Store Hollofil II Sleeping Bags

Bags filled with Dacron polyester Hollofil II have excellent recovery after compression. However, storing while compressed can, in time, reduce the recovery. For best results, air the bag, fold it, and store it. Roll it or put it in a stuff sack *only* to transport it.

Garments insulated with Dacron polyester Hollofil II should be handled the same as sleeping bags.

THE SYNTHETIC FILLS/CELANESE POLARGUARD

Polarguard is the only continuous filament polyester fiberfill for cold-weather insulation. It is manufactured by a unique threaded roll processing system. Polarguard is lightly bonded with acrylic resin, and heat-cured to insure dimensional stability with use and laundering.

The continuous filament batting provides superior strength and lasting loft. It has a definite advantage, we feel, over DuPont's Hollofil II in the manufacture of sleeping bags. The difference between the two fibers is that Polarguard is long-stranded, and prone to less shifting within a sleeping bag system. Another advantage of Polarguard's continuous filament construction is that sleeping bags made with this fill need *much* less quilting than other bags.

Polarguard absorbs only 1 percent of moisture—a real advantage over down. It is non-allergenic, odor-free, and cannot be attacked by insects or microorganisms, including mildew.

Polarguard sleeping bags have been extensively field-tested, with no substantial loss of loft or insulating qualities.

HOW TO CLEAN AND MAINTAIN SLEEPING BAGS INSULATED WITH POLARGUARD

Follow the instructions previously listed for the care of sleeping bags insulated with Hollofil II.

The Simple Sack—a Synthetic Fill
(Photo courtesy of Camp Trails Company)

HOW TO CARE FOR AND MAINTAIN DOWN SLEEPING BAGS

Down* is the lightest insulating material known. It is the best fill that can be used for clothing or sleeping bags because it can compress to less than an eighth of its potential loft, it weighs almost nothing, and it's the warmest insulation per square inch. Its liabilities are that when it gets wet, it totally loses its lofting abilities and insulating properties. Some people are also allergic to it. In addition, it's very expensive: As we write this book, a 2 ¼ -pound prime Polish goose down sleeping bag is selling for over $250.

In order to maintain the loft of your down bag and extend its life expectancy, you must keep it clean. While on the trail, you *must* give your sleeping bag a chance to air out, so that perspiration and moisture that have been absorbed by the down pods can evap-

*There are many different qualities of down. The best quality down, generally, is Polish goose down. Any garment filled with this material will be *marked* "Polish goose down." The next best quality goose down is "prime" or "AAA" goose down; this means that 100 percent of the fill is goose down. This goose down may be Chinese in origin. Anything else can be labeled "down." Down can consist of any quantity of goose down blended together with any quantity of duck down. Duck down has less lofting capability, is usually not as warm as goose down, and doesn't have goose down's resiliency. It's also cheaper.

orate. This is best accomplished by opening the zipper to its full length, turning the bag inside out, and placing it on top of your tent (make sure the tent has no morning dew on it). If the tent is wet, string a line between two trees and suspend your sleeping bag from it. Avoid prolonged exposure to the sun, as the sun can leech out the natural oils in the bag.

How to Hand Wash a Down Sleeping Bag

We do not recommend hand washing your prized down sleeping bag. We've done it ourselves, with good results, but *extreme care must be taken when removing the wet bag from your bathtub;* the weight of the excess water retained in the baffle sections of the bag can cause the baffles to rip internally. This is a *very* expensive thing to repair.

Some of you may not take our advice and decide that you want to wash your own down bag. Here's how. Fill your bathtub half full with lukewarm water and add either Ivory Snow, Woolite, or Nu-Down. Either half a cup of Ivory Snow or four or five capfuls of Woolite will be sufficient. If you're using Nu-Down follow instructions on the package. *Do not under any circumstances use a detergent.* Disperse the soap through the water with your hands.

Starting with the foot section of the bag compress the lower baffles in your hands, immerse them in the tub, and then release that section of the bag. Continue this process until the whole bag is immersed.

Knead the bag gently with your hands. Let the bag soak about fifteen minutes. Then turn it inside out and repeat the process.

Drain the tub while the wet bag is still in it. Refill the tub halfway with lukewarm water, and again the prescribed amount of soap. Knead it through the bag again.

Drain the tub and gently compress all water out of the bag.

Fill the tub halfway with lukewarm water and rinse the bag out by gently compressing the baffles and swishing the water around, then drain the water, repeat this process once more. You *must* give the bag at least two rinsings in lukewarm water.

Very gently remove the sleeping bag from the tub and spread it out on thick towels. Put towels on top of the bag and inside the bag to absorb as much water as possible.

Your sleeping bag will now look a sorry sight, a lumpy stretch of nylon paneling. Break up as many of the clumps of down as you can by moving them between your fingers. Then take your bag to an oversized, front-loading commercial dryer. *Do not* put it in your home dryer, which will be too small and will damage your bag. Place a *clean* sneaker in the dryer with the bag. Set the dryer on the *lowest possible temperature*. Take at least three or four dollars worth of dimes with you—and a good book; you'll be there from two to four hours. *Do not* rush the process by increasing the heat of the dryer. Every time the dryer stops (don't put in more than two dimes at a time), check the bag for clumps of down that may be forming in the panels. Break up these clumps by hand and return the bag to the dryer.

You'll know the bag is dry when it returns to its natural loft and there are no clumps. Transport the bag back to your house in its storage bag—*not* in its stuff sack. When you get home, take it out of the bag and leave it open for 24 hours.

Professional Cleaning for Down Bags

Do *not* take your down bag to a local cleaning store; most cleaners simply don't know how to care for down-filled garments and sleeping bags. Ask your local backpacking retailer if he can recommend a down cleaning service. If not, there are two specialists with whom we've dealt who accept down cleaning by mail. Both are reliable and do excellent work. Write for a price list first, enclosing a self-addressed stamped envelope:

The Down Depot
108 A Carl Street
San Francisco, California 94117

Down East
93 Spring Street
New York, New York 10012

How to Store a Down Sleeping Bag

Never store your down sleeping bag in the stuff sack. We recommend that you hang your sleeping bag so that it can remain fully fluffed when not on the trail. The motto at Down Depot is "Hang loose when not in use." If this is not possible, make or purchase a large oversize stuff sack of non-waterproof material. Store your sleeping bag in this package.

How to Waterproof a Down Sleeping Bag

The only way to waterproof a down sleeping bag is to buy a Gore-Tex® cover for it. The three best Gore-Tex® covers on the market at the time of this writing include: the Early Winters, Ltd. Bivvy Cover, the Blue Puma Bare Necessity, and the Marmot Mountain Works Burrow. It's probably going to be a while before these covers appear at sales or in bargain basements, so you may have to compromise and pay list price: a pretty darn steep fifty to one hundred dollars. It's inevitable that more manufacturers will begin

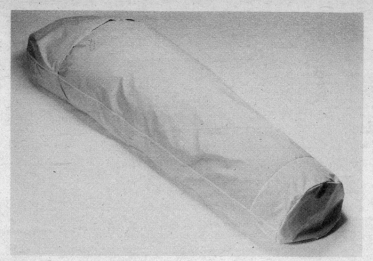

Early Winters Gore-Tex® Bivvy Cover
(Photo courtesy of Early Winters, Ltd.)

producing these bags, and that they'll start appearing at sales and in clearance centers—so you may just want to wait it out. We've ordered ours now.

HOW TO REPAIR A TORN OR DAMAGED SLEEPING BAG

It's inevitable that you're going to get at least one little tear in your sleeping bag. Carry ripstop nylon repair tape with you and patch the damage immediately. In an emergency, adhesive tape will work, though it leaves an awful resin. Be sure to sew the ripstop nylon tape to the bag as soon as you get home.

Major damages resulting from burns or other causes are best dealt with by a reliable repair service.

SUPER-BUDGET ALTERNATIVE SLEEPING BAGS

G.I. Mummy Bags

The G.I. mummy bag was generally filled with feathers and duck down. You can still pick up one of these bags in excellent condition for between $20 and $40 at most military surplus stores. Anna used one of these bags for about six years, and it was a great three-season bag. Optional military covers are still available for these bags, priced between $2.00–$4.00. This outside cover adds additional insulation and wind resistance to the bag, but it also adds too much additional weight. These bags are a good investment, especially for young people who are very hard on their equipment; the bags have been around almost thirty years already, and they've proved virtually indestructible; they're easily repaired with a needle and thread. At present, the Army is using goose down bags in such places as Antarctica and Greenland; this new equipment occasionally does show up on the military surplus store circuit. If you do find a goose down G.I. bag, rest assured that it's of superior quality. As an example of the quality of goose down gear the Army has been ordering, two years ago we found a pair of modern U.S. Army surplus goose down bib overalls for $10. Right inside them we found a very well-known label: Eddie Bauer! We don't mind the Army's khaki color—not if we can pay $10 for Bauer down overalls!

BLANKET ROLLS

There is no reason now why a beginner backpacker can't use a blanket roll. To make a blanket roll, follow the diagram on page 75.

How to Fold Blankets for a Blanket Roll

You'll need two blankets and four blanket pins.

Place one blanket flat on the floor. Place the second blanket on the floor so that it overlaps half of the first blanket. Refer to the diagram showing sections A, B, and C.

Fold Section A over Section B. Fold Section C over Sections A and B. You now have an even rectangular tube with four layers of blanket.

Fold one end of the blanket roll up 6 inches and pin it with two blanket pins. At the other end pin the top two blanket layers together about 6 inches from the edge. And then pin the bottom two layers of blanket together. When evening comes you slip between the blanket layers.

HOW TO CONVERT YOUR SUMMER SLEEPING BAG INTO A WINTER BAG

If you own a down three-season bag which you've found isn't warm enough for chilly winter excursions, you can give your bag

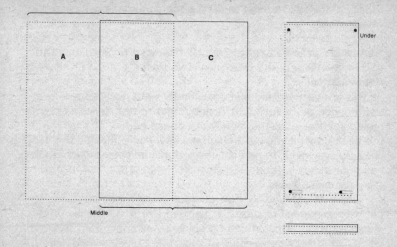

added warmth with a one-pound down liner. You'll probably pay up to $65 unless you can find one on sale. Unfortunately these liners rarely appear on sale, but ultimately it will cost you less to buy one at full retail price than it will to buy a winter bag. Holubar offers a down liner kit for those who want to sew their own.

Another way of warming up a lightweight sleeping bag is to take an extra blanket with you—and put it inside the bag (as long as extra weight is not a problem).

If you have a summer sleeping bag that you want to adapt for winter use, here's another idea. Buy an overbag for about $35 to $45. When you shop for an overbag, look for Polarguard. We personally are very satisfied with our Snow Lion Polarguard overbag, and right now we're keeping our eyes open for another one either on sale or in one of our local bargain basements.

SOME GOOD VALUES FOR THE MONEY/SLEEPING BAGS
REI McKinley

This is one of the best buys available on a pre-sewn down bag. It costs under $100. It's suitable for temperatures just below 0° Fahrenheit, and it's the most popular sleeping bag that REI produces. The McKinley has a 40-inch long central zipper, and 6 inches of loft. The regular model is 86 inches long, and is filled with 2½ pounds of prime goose down. Total weight is 4 pounds. The only liability of this bag is that it can't be zippered together with another sleeping bag.

REI Little Takoma

A child's sleeping bag of Hollofil II, this bag is priced under $40. A nice little bag for the wee ones in your life.

REI Mount Baker

This is a summer bag that costs under $40. It's filled with over 2½ pounds of Hollofil II. It opens up to form a quilt, and is a remarkable value for the money. An added benefit of buying REI products is that if you are a co-op member (initial membership fee is $2.00), you get a 10 percent equipment credit at the end of each year. More details about REI appear on page 28.

Greylock Grand Teton

Filled with 5 pounds of Polarguard, this is the warmest sleeping bag Greylock manufactures. Its finished size is 31 inches × 94 inches. This bag also features a 68-inch 2-way zipper. Here is a fine value and a really good bag that's marketed under several labels. The price will vary between $65 and $95. Obviously, it's a better buy at $65, but it's still a good value at $95.

Coleman Peak I, Model #740

A new sleeping bag in the Coleman line, this model is made with Dacron II fill and it weighs 5 pounds 12 ounces. It features a full length center zipper and drawstring hood. The suggested retail price is under $60.

Co-op Wilderness Supply Sierra Sack

Polarguard filled, this bag weighs 3 pounds 4 ounces. It is made of 1.9-ounce ripstop nylon, and it has a 72-inch 2-way side zipper that will mate with another Sierra Sack. It is serviceable in the temperatures above 20° F. It costs under $60.

Moonstone Mountaineering Sleeping Bag Lightweight Model

Moonstone Mountaineering is a small independent manufacturing concern specializing in Polarguard-filled products. We learned of this sleeping bag only recently, and were very impressed by its construction and design. The stitching and finish are excellent. It's a semi-mummy design, with the hood section coupled to insure a more comfortable fit when the hood is drawn tight. The baffle over the zipper contains the most generous amount of Polarguard we've seen. It has 5 inches of total loft, and weighs 3 pounds 8 ounces. A very attractive and comfortable bag, it lists for around $80. If you're over 6 feet 4 inches, Moonstone will custom construct a bag for you at a reasonable price.

EMS Berkshire

This mummy-style bag with drawstring hood has a three-season rating of 25° F. Total weight of this Polarguard sleeping bag with a 4½-inch loft is 3 pounds 15 ounces. It is sized to fit someone up to 6 feet 2 inches, and costs under $60.

EMS Blueridge

This is a 4-pound, semirectangular bag good for summer backpacking. It has a full-length zipper which can be opened to form a quilt. It's sized to fit backpackers as tall as 6 feet 1 inch. It's priced under $50.

The EMS Pocono

This is a child's sleeping bag, (children up to 5 feet tall). It has 4 inches of loft and opens up into a quilt. This summer sleeping bag costs less than $40.

Store Brands

If any of the bags you see are store name brands, ask the retailer who manufactured them. Sometimes you'll be surprised to find out that Snow Lion, or Trail Tech, or Hunter Sleeping Bags/Greylock Industries was actually the producer of the bag. For example, the EMS Franconia and Berkshire (the heaviest weight) are produced by Snow Lion and are great values for the money. The lightweight Berkshire was produced by Trail Tech—not quite as well-made as Snow Lion products, but still a good value. Also bear in mind when comparison shopping that there are only a few ways to manufacture synthetic-fill sleeping bags. Don't pay extra for a fancy manufacturer's label.

SLEEPING BAG KITS

The cheapest way to buy a new sleeping bag is to buy a kit and make it yourself. There are several kits on the market we can recommend. Each will save you from 35 to 50 percent off the price of equivalent bags.

Frostline Big Horn

Priced under $100 this bag is the equivalent of a $170 down-filled winter mountaineering bag. It's an excellent design, though it's not recommended as a project for someone who has never assembled a kit.

Frostline Nimbus

Made with Polarguard fill this kit is priced around $50. An excellent spring and summer sleeping bag, it is well worth the money.

Altra Kits Polarguard Sleeping Bag #235

This model retails for less than $50. Altra kits, available at many backpacking equipment retail shops, come with very easy-to-follow instructions.

EMS Kit Seneca Hollofil Sleeping Bag

An oversized mummy bag that's reasonably priced (under $50), this bag is also well designed. Unfortunately, it weighs six pounds —the heaviest sleeping bag we've seen from a kit.

Holubar Polar Sleeping Bag Kit

Holubar makes the Cadillac of sleeping bag kits. The price is just over $100, but a comparable pre-assembled bag would cost you $200. For $30 extra, it's possible to purchase extra down fill for this bag to make it an expedition-grade bag.

5

HOW TO BUY A NEW PARKA OR VEST

———————————————— **About Parkas and Vests**

For most backpacking situations, a warm sweater, a down- or synthetic-filled vest, and a wind shell will suffice. Wearing a woolen sweater and a parka will eliminate the necessity of having a wind shell. Do *not* buy a parka that is designed for winter mountaineering or expedition use unless you are, in fact, going to use it just for winter mountaineering; if you use it in any season but winter you'll find it much too warm.

Parkas can be quite costly although the prices will vary widely and fluctuate with the season. Keep your eye open for summer sales.

ADVANTAGES AND DISADVANTAGES OF A DOWN-FILLED PARKA

A down-filled parka is very lightweight and extremely warm. It is also easily compressed into a stuff sack, or into a pack side pocket. When packing for a four- to five-day (or longer) backpacking trip, even in the summer, it makes good sense to have a down-filled parka.

There are disadvantages to down parkas, too. Like down-filled sleeping bags, they are very expensive—and they are not practical in wet weather. If it rains, you could wind up wearing a soggy mess. You really must carry a rainjacket if you plan to use a down parka. There are now some parkas covered with Gore-Tex,® a shell that would eliminate your having to carry a separate rain jacket. But Gore-Tex® is expensive too.

ADVANTAGES OF SYNTHETIC-FILL PARKAS

Polarguard and Hollofil vests and parkas are much cheaper than similar down garments. They're a little bit heavier in weight, and they don't pack as easily as down, but they can be just as warm.

Additionally, if it rains and the garment gets wet, it can be wrung out and won't lose most of its loft. Nevertheless, we advise you to use a rain shell over these parkas, too: No one wants to walk around in clammy wet clothes.

HOW TO DECIDE WHAT LENGTH YOUR PARKA OR VEST SHOULD BE

Stand up straight with your hands at your sides: You should be able to touch the hem of your parka with the tips of your fingers. We caution you against buying a shorter parka; it doesn't give adequate protection to the kidneys.

A vest should reach below your waist. It is designed to be worn over a woolen sweater in mild weather, or under the parka in colder climes. Some vests come with an extra long back that covers the kidneys; this is a desirable feature to look for. Eddie Bauer used to make the finest down vest available—the Snow Line Vest. It was the longest vest on the market. Unfortunately this model has been discontinued, but perhaps you'll find one secondhand. We think an astute manufacturer should take the hint and start to produce another long vest.

HOW TO DECIDE WHAT WEIGHT PARKA OR VEST YOU NEED

Weight should be defined in terms of loft. The warmth of a parka has everything to do with how thick it is and how much dead air space is enclosed within. From an 1 to 1½ inches of loft is perfectly adequate for spring, summer, and fall backpacking.

When you are sizing your jacket or vest, be sure to try it on with a medium-weight sweater so that it won't be too tight when you need to wear the extra garment for warmth.

If you're planning to go winter backpacking, you'll need a winter parka, which should have between 2 and 3½ inches of loft. A hood, of course, is *essential* for a winter parka.

THE DIFFERENT CLOSURES

There are three methods of closing vests and parkas: snaps, zippers, and Velcro. We think that the best system for a backpacking parka is one that employs a two-way zipper with a draft tube over it that closes with either snaps or Velcro.

Jansport Mountain Vest
(Photo courtesy of Jansport)

Our own preference for vest closures is snaps. They weigh less than a zipper and are less messy than Velcro. (Velcro tends to pick up lint, hair, and dust). But any closure system really will work well on a vest.

WHAT TO LOOK FOR IN CONSTRUCTION DETAILS

You will want a collar that extends up your neck 1½ inches. You'll also want to be sure the zipper is covered by a draft tube. Deep square-cut pockets that are roomy enough for gloves, lunch, or anything else you may want to carry within easy reach are a necessity. You should look for flaps over the pocket, and a separate hand warmer pocket behind the regular pocket. Check the stitching around the pockets to see how they are attached to the body of the parka. Also check the stitching to see that the zipper has been attached well.

Check for any thin spots under the arms of the parka; some manufacturers have produced parkas with almost no insulation in this sensitive area. If you find inadequate insulation under the arm, don't buy the parka. Also, if you're about to buy a down parka, try on several of the same parkas. Sometimes the very same model by the same manufacturer will have less or more down filling than the others you find on the same rack. Buy the one with the best loft, that is the most insulation and thickness.

A parka with a snap-on hood may be desirable for backpacking in cool weather; by combining it with a heavy woolen sweater you can extend its range of use by about 15 to 20 degrees.

A parka should have a tightening device inside which will prevent a cool breeze from entering the bottom of the parka. A two-way zipper aids in ventilation and in maintaining proper body heat when hiking with a pack.

Make sure that if elastic is used in the wrist closures, it's not too tight. Even moderately tight elastic can affect proper circulation of the blood.

Lastly, more and more parka and vest construction has been shifted to Hong Kong and Taiwan—and the manufacturers are not passing on to you the amount saved by use of cheap labor. If you have a choice, we recommend buying the parkas produced in the United States: Some well-known manufacturers are Holubar, Gerry, Marmot Mountain Works, Sierra Designs, North Face.

SHOULD YOU BUY A NAME BRAND PARKA OR VEST?

It is not necessary to buy a name brand parka or vest in order to get warm, well-made, quality garments. However, be wary of parkas and vests made in Hong Kong, Taiwan, and Formosa. By now you've probably heard about the chicken feather scandal; these countries were known to export garments filled with chicken feathers. That doesn't mean that there aren't any in some American-made "down" garments.

HOW TO STORE A PARKA OR VEST

Do *not* leave your parka stuffed into its stuff sack. Hang it on a well-constructed wooden hanger in a well-ventilated closet, away from the sun. Never hang up your parka where there is constant sunlight. Treat your vest the same way.

If you're going to put your parka and vest away for any length of time, be sure to wash or clean them first.

SOME GOOD VALUES FOR THE MONEY/PARKAS
REI Glacier Parka

This lightweight, all-purpose parka has an outer shell of ripstop nylon. It's filled with Hollofil II and it has a loft of two inches. The Glacier Parka is 29 inches long. It has a two-way zipper, hand-warmer pockets, and elastic cuffs. This parka comes with a hood, and costs slightly over $50.00.

Co-op Wilderness Parka

This very warm 28-ounce parka is filled with twelve ounces of down. It has a two-way zipper, with a snap wind tube, elastic cuffs and two cargo pockets that close with Velcro. It is priced a little over $50.

International Mountain Equipment Polarguard Parka

This parka, manufactured by Snow Lion, is filled with 18 ounces of Polarguard, and is covered with 1.9-ounce ripstop nylon. There are no sewn-through seams, which reduces wind penetration. A two-way zipper, a snapped wind tube, two large cargo pockets, as well as handwarmer pockets, are some other features of this parka. It retails at less than $50.

Snow Lion Palisades Parka

This parka has double-quilt Polarguard construction, cargo pockets, handwarmer pockets, a two-way zipper, and a generous wind tube. This parka is virtually identical to the IME Polarguard Parka listed above, although it sells for three or four dollars more.

International Mountain Equipment Polarguard® Parka
(Photo courtesy of International Mountain Equipment)

SOME GOOD VALUES FOR THE MONEY/VESTS

International Mountain Equipment Polarguard® Vest

A double-quilted vest made from two offset layers of Polarguard, this garment weighs 15 ounces. The shell is 1.9-ounce ripstop nylon. The vest has very generous cargo pockets, a three-inch high collar and no sewn-through seams. It costs less than $30.

Trailwise Columbia Vest

A very simple ripstop nylon shell filled with Dacron II, this vest has sewn-through seam construction, snap closures, handwarmer pockets, and a high wind collar. It costs a little over $25.

Co-Op Down Vest

Filled with over six ounces of quality goose down, and designed for rugged use, this vest is one of the best values on the market. It has two slash pockets and a zipper front, and it is two inches longer in the back. It, too, costs about $25.

EMS Greylock Vest

This vest is available in men's, women's, and children's sizes. It is filled with ten ounces of Polarguard; it has snap closures, hand-warmer pockets, and a very high collar. The most attractively styled vest we've mentioned, it retails under $25.

PARKA AND VEST KITS

Parka and vest kits will save you 40 to 50 percent of the cost of a comparable new, ready-made garment. Some of the kits for down sweaters and vests are easy even for the beginning sewer.

Country Ways Kits

Country Ways produces a very serviceable Polarguard vest with an average loft of 1½ inches. The medium-size vest is 31 inches long in the back, and will fit anyone with a chest measurement of 39 to 41 inches. The model number is A101. It has a nylon taffeta outer cover and inner lining. It also has two very deep hand-warmer pockets. It's a very durable, easy-to-construct vest.

Holubar Sew-It-Yourself Kits

Holubar produces a really outstanding goose down vest with optional zip-on sleeves. It has the highest collar of any down vest manufactured in the United States. It's an outstanding design that features a snap front, Velcro-covered storage pockets, and hand-warmer pockets. It will cost you a little over $20.00. The Model number is 1164.

The Holubar Polarguard Parka, kit number 1301, is a really rough and tough knockabout parka. It has a 5-inch collar, zipper and snap closures, roomy cargo pockets that seal with Velcro, as well as handwarmer pockets. It also has an inside drawstring. The outer fabric is 1.9-ounce ripstop nylon. It is an excellent buy.

Frostline Kits

The Tundra Jacket, model number C10M, is 33 inches long and comes in men's and women's sizes. Made of prime goosedown, this is probably the warmest parka you can make for the money. It's not the most flattering design, but if you're interested in warmth and savings rather than style, this is the parka kit to get. It costs less than $50.

Eastern Mountain Sports Kits

EMS Hollofil Vest Kit number 820944 is the least expensive vest you can make (cost is under $15.00). The vest is covered with ripstop nylon that comes in several colors. There is a zipper opening and the back is three inches longer than the front. This is an excellent choice if you're making a vest for a child.

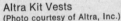

Altra Kit Vests
(Photo courtesy of Altra, Inc.)

Calico Kits

The North Country Down Jacket is made of 1.9-ounce ripstop nylon, and is filled with 10 ounces of prime goose down. It has two very deep cargo pockets, handwarmer pockets, and elastic cuffs with tightening snaps. This jacket is nicely styled and comes in both children's and adult sizes. The adult size costs under $45.00, and the children's size costs about ten dollars less.

6

HOW NOT TO GET BURNED BUYING A BACKPACK STOVE

————————————————————————**About Stoves**

Buying a backpacking stove used to be a small investment. One of the first gasoline backpacking stoves that we used was the Svea #123; it cost $6.00 in 1969. The Svea 123R, basically the same stove, but with a self-cleaning needle, now costs around $30.

In 1975, we purchased three Optimus #111B stoves for use by our climbing school, North Country Mountaineering. We paid $25 for each stove. The Optimus 111B now retails for $55! There is no reason to believe that these crazy prices are about to stop rising.

Obviously, most backpackers who are planning to spend one or more nights in the woods will want to have a reliable stove with them. This chapter will show you how you can do that without going broke.

ADVANTAGES OF THE GASOLINE STOVE

It is a regulation in many Federal and state parks that all cooking be done with portable stoves. No open flames or fires are permitted. We feel that a gasoline stove is the most practical stove for use in the American woodlands. Fuel can be purchased at most hardware stores, camping supply stores, and outdoor outfitters. This fuel (Coleman or white gas) is inexpensive, from $1.85 to $2.95 per gallon; consequently, this stove is a good choice for the budget-minded backpacker.

ADVANTAGES OF THE PROPANE AND/OR BUTANE STOVE

The main advantage of the propane and/or butane stove is that you don't have to handle any flammable mixtures. The fuel comes in pressurized steel containers, and refueling is very simple. Stoves using these fuels have several notable disadvantages though. The

Optimus 111B
(Photo courtesy of Optimus International)

canisters are affected by cold and may not function reliably on winter backpacking trips. Another problem is that it's extremely difficult to tell how much fuel is in the canister, and it's dangerous to change if it still has fuel left in it; this means that if you have a partially filled canister attached to your stove and you're about to go out on an overnight trip, you still have to carry both the partially filled canister and a new one.

There are two other disadvantages: Empty canisters must be packed in and out, and propane/butane fuel is getting expensive; a three-day, two-night backpack may cost you $4.00 just for fuel.

Bleuet Globetrotter Butane Stove
(Photo courtesy of Wonder Corporation of America)

ADVANTAGES OF THE KEROSENE STOVE

Kerosene is probably the cheapest fuel you can buy. It's available worldwide—a real advantage for backpackers who are going abroad. But it, too, does have disadvantages. It smells bad, it tends to smoke (incomplete burning), and it can leave a funny taste in your food if combustion is not complete.

ADVANTAGES OF THE ALCOHOL STOVE

Alcohol does burn slowly, it's generally less flammable than other fuels, and it's most often used on boats. These advantages must be weighed against the fact that it is probably the most expensive fuel to use. And denatured alcohol is hard to find.

SOLAR COOKING

This is a totally new concept in outdoor cooking, and it's definitely an idea whose time has come. The Sunpack Solar Cooker by E-Z Sales and Manufacturing of Gardena, California, is inexpensive to buy, is relatively easy to set up and maintain, and requires no fuel other than sunlight. Steven tried out a prototype of this device at a recent Appalachian Mountain Club Inter-Chapter Mountaineering meeting and drew quite a crowd with it. While it's true that you have to turn the solar cooker as the direction of the sun changes, it's still novel enough and a sufficiently appealing concept to warrant the extra bit of attention and effort. This wonderful device provides an effective method for cooking, boiling, or frying in minutes between 10 A.M. and 4 P.M. on sunny days. It cooks summer or winter, at any altitude, on any day when the sun casts a shadow. The Sunpak Solar Cooker can also be used as a campfire reflector oven. Just be careful when setting it up not to cut your hands on the sharp panel edges. Also, when you're out in the sun and cooking, do *not* stare at the reflector; you could damage your eyes. We recommend that you use sunglasses when using this device.

COOKING WITH STERNO

Those of you who are going backpacking for the first time might consider the purchase of a collapsible Sterno stove. The stove itself runs about $2.50 to $3.50 at most discount department stores, and at hardware stores. The Sterno fuel cans cost about 75¢ each. We used Sterno for some time, particularly during our Scout days. It's a perfectly viable way of heating your evening meal.

HOW TO SELECT THE RIGHT STOVE FOR YOUR NEEDS

If you are caught in a rainstorm, and you're cold, wet, and hungry, you want a stove that's going to work when you light it. Once you have decided what kind of fuel you want to use, the following questionnaire will help you decide what kind of stove you want— and what kind of stove you need.

1. Will I be backpacking by myself, or with one other person?

If you're backpacking with one other person, you need a stove that will allow fast cooking in medium-size pots. You can get a slightly heavier stove knowing that your backpacking partner can carry the fuel. The Optimus 8R, the Optimus 99, and the Bleuet S200 are the best 2-person backpacking stoves you can buy.

2. Will I be backpacking with two other people?

For backpacking parties of three we recommend the Phoebus

725, the MSR gasoline stove, and the Coleman Peak I. These are heavy duty stoves capable of very fast cooking. The Phoebus and the Peak I are especially stable for large pots. The Peak I and MSR stoves have built-in pressure pumps for ease of lighting. The MSR is the most packable and collapsible of the three. The MSR can also be used handily as a one- or two-person stove.

3. Will I be backpacking with three other people?

The best stoves to use for cooking for a group of four people are the Optimus 111B, the Optimus 00, the Optimus 111 (kerosene) and the Phoebus 625. These are the only four stoves which are capable enough and dependable enough to be used as the sole cooking source for a group of four people. The Optimus stove is extremely stable. All the stoves mentioned feature a built-in cleaning needle, a built-in pressure pump, stability for large pots, and large-capacity fuel tanks.

Although the stoves we mentioned will serve the needs of four backpackers, we advise a group of this size to use two stoves. This allows for the cooking of two separate items at the same time. For an extended trip, when there's bound to be a lot of cooking, a super combination is an Optimus 111B, and a MSR stove. This requires an investment, however, of at least $100. We have used two Peak

MSR Multi-Fuel Stove
(Photo courtesy of Mountain Safety Research)

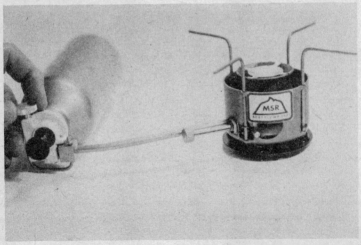

1's very successfully on such a trip, and were thus able to keep the cost under $55.

4. Will I be backpacking with four or five other people?

With a group of six, you *must* have two stoves. Two Coleman Peak 1's would do quite well. Two Optimus 111B's would be even better. Three stoves would permit two dishes to be cooked at the same time, plus a pot of water for coffee or dishwashing.

5. Do I want a pressure pump?

To avoid spilling gasoline on a stove when priming it, a pressure pump is fantastic. All stoves that come with a pressure pump are self-priming. To ignite, all one has to do is increase pressure by priming the pump, turn the switch, and light a match. If you're thinking of buying the Svea, the Optimus 8R, or the Optimus 99, you can buy an optional minipump which aids in starting these little stoves and avoids any liquid priming.

Phoebus 625 Camp Stove
(Photo courtesy of Precise Imports Corporation)

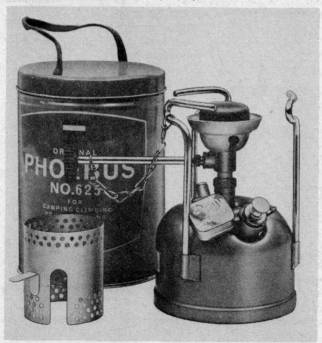

6. Do I want a stove that needs priming?

We ourselves choose not to have any stove that requires liquid priming. However, if you decide to buy a stove that does need priming, we recommend you use a flammable fire paste such as Mautz Fire Ribbon, or Optimus Burning Paste. These pastes, when ignited on the stove, produce heat that facilitates priming.

7. Do I want a stove for large pots?

The stoves best designed for large pots are: the Optimus 111B, the Optimus 111, the Phoebus 625, the Phoebus 725, and the Coleman Peak 1.

8. Do I want a stove for winter use?

The best and most expensive winter stoves are the Optimus 111, the Optimus 111B, the Phoebus 625, and the MSR stove. Good quality winter stoves are: the Optimus 00, the Coleman Peak 1, and the Phoebus 725. With the addition of an Optimus minipump, the Optimus 8R and the Optimus 99 stoves are also acceptable for winter use. Winter backpackers require stoves that function reliably and efficiently. A stove that does not work can spell the difference between a successful trip and disaster. Maintenance is particularly important for the winter stove. Before setting out on a winter trip, be sure to take your stove apart and clean it.

We recommend that you cut a piece of ¼-inch plywood a little wider than the size of the base of your stove, and place it under your stove while cooking. This will prevent heat transfer onto tent fabric or snow.

9. Do I want a stove for just spring, summer, and fall backpacking?

For spring, summer, and fall backpacking, gas stoves and propane and butane stoves are very handy. We've been very satisfied with the simplicity of the butane-powered Globetrotter Stove by Bleuet. It's very compact, and extremely lightweight (even with the fuel canister), but expensive to run. An alternative butane stove you might want to consider is the Bleuet S200. The fuel cartridge it uses will last twice as long as the Globetrotter's at about a 33 percent savings. Here are our favorite stoves for three-season backpacking: When Steve is hiking alone, he takes a Svea 123. This is one of the lightest of the gas stoves, and has a convenient built-in cleaning needle. Anna owns a second-hand Bleuet S200, so that's what she uses. For two-person backpacks the Optimus 8R and the Optimus 99 are highly recommended.

10. Am I willing to carry a little extra weight for a stove that works better?

If you've answered "yes" to this question, then you think the way we do. When we hike with another person or in larger groups, we usually distribute the load—so the extra ounces aren't noticed. For example, we like to take the Optimus 8R with the minipump along during fall season excursions; Steven is always brewing up, and the 8R is one of the few stoves that has a capacity that matches his.

BUYING A STOVE ON SALE

Most of the leading outdoor chain stores and specialty outdoor shops hold an annual sale, at which time you should be able to buy a stove at 25 percent off the regular list price.

If you can't wait for the sale, ask the manager of your local store if any floor models are for sale; these will have been used in the store for demonstration purposes only, and you ought to be able to get one for from 15 to 25 percent discount off the regular price.

SOME GOOD VALUES FOR THE MONEY/STOVES
Bleuet S 200

This stove costs under $15. It's a reliable, dependable, spring-summer-fall stove and is very easy to operate. The stove weighs one pound. We suggest the use of a wind screen with it.

Coleman Stove
(Photo courtesy of
the Coleman Company)

TABLE OF SPECIFIC DETAILS FOR STOVES

	Weight Complete (oz.)	Weight w/fuel (oz.)	Dimensions – Height (in.)	Width/diam. (in.)	Length (in.)	Fuel
Bleuet S-200	15.0	24.0	4½	3½	9½	Butane
Globe Trotter	17.0	23.0	5⅝	4⅛	—	Butane
EFI Mini	8.0	17.0	—	4½	1½	LP Gas
Optimus Mousetrap	12.0	22.0	1¾	4¾	7	LP Gas
Optimus 00	26.0	39.0	4½	5½	7	Kerosene
Svea 123R	17.5	22.0	5	4½	—	White Gas
Optimus 8R	23.0	26.0	3¼	5	5	White Gas
Optimus 99	23.0	26.0	3	4½	4½	White Gas
Optimus 111B/111	56.0	68.0	4	6¾	7	Wht. Gas/Kero.
Phoebus 725	23.0	31.0	4½	5½	—	White Gas
Phoebus 625	32.0	48.0	7¼	5½	—.	White Gas
MSR Gasoline	15.0	32.0*	3½	3¼	9¾	White Gas
Peak 1	31.0	41.0	6½	4⅝	—	White Gas

* with 16 oz. of fuel
ᴵ Boiling time is time it took to boil 1 quart of average temperature tap water.
The stove was operated in a warm room. Times will vary greatly with
altitude, air temperature and wind.

TROUBLE SHOOTING STOVES

Please refer to the exploded view of the SVEA 123R.
All the gasoline and kerosene stoves we sell are covered by the principles in this guide.

Problem	Possible Cause	Solution
Failure to start	Pre-heating insufficient, thus pressure insufficient.	Fill burner head to overflowing and reignite.
Failure to operate	Leaded gasoline Vapor leak Clogged burner tip	White gasoline only Tighten at points A, B, C, or D Clean tip NOTE: Remove burner tip from stove before cleaning unless stove has internal cleaning needle.
	Impacted debris from long use or carbon (caused by overheat) in cotton wick inside vaporizing tube.	Replace wick — major job, best done at home.
Gas or flame leak at point B	Packing nut loose Packing worn Spindle worn or broken	Tighten Replace Replace
Gas or flame leak at point C	Vaporizer tube loose at base	Tighten
Gas or flame leak at point D	Tank cap gasket missing, cracked or burned	Replace
	Internal plug in plunger askew	Unscrew top of tank cap and reseat rubber plug in plunger. NOTE: When assembling, be sure cap is firmly tightened yet can be easily loosened again.

Fuel Capacity (oz.)	Boiling Time¹ (min.)	Burning Time² at Simmer (min.)	Pressure Pump	Built-in Cleaner	Needs Priming Fuel	Cold Weather Usage	Simplicity
6.3	5	190	NA	NA	No	Poor to No	High
3.2	10	80	NA	NA	No	Poor to No	High
6.2	6	195	NA	NA	No	Maybe	High
6.2	8	185	NA	NA	No	Maybe	High
13	4½	90	Yes	No	Yes	Yes	Low
6	6	60	No	No	No	Maybe³	Avg.
3.2	7	70	No	Yes	No	Maybe³	Avg.
3.2	7	70	No	Yes	No	Maybe³	Avg.
16	6	110	Yes	Yes	No/Yes	Yes	Avg.
10	7	105	No	Yes	No	Yes	Avg.
16	4½	190	Yes	Yes	No	Yes	Avg.
16 or 32	4	130*	Yes	No	No	Yes	Avg.
10	3½	210	Yes	Yes	No/Yes	Yes	Avg.

² Burning time is total time stove can maintain a rolling boil in a quart of water. These times will increase if the stove is set to a light simmer, and decrease if the stove is opened to its maximum.

³ Stove will operate well in winter if the Optimus Mini-pump is used.

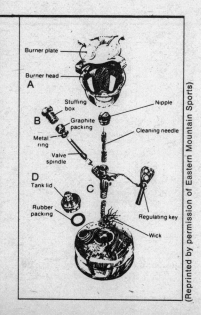

Burner plate
Burner head
A
Stuffing box
Nipple
B
Graphite packing
Metal ring
Cleaning needle
Valve spindle
D
Tank lid
C
Rubber packing
Regulating key
Wick

Coleman Peak 1

Besides being the quietest of all gasoline stoves, the Coleman Peak 1 features a built-in pump, and built-in cleaning needle. It's particularly stable for large pots. It weighs 1 pound 15 ounces, and costs about $30. This may seem like a big investment, but it's a large capacity, four-season stove. It's also the cheapest stove with a built-in pump.

Bleuet Globetrotter

This cartridge stove folds into a compact unit 3½ inches high, 4⅛-inch diameter and that's with the fuel cartridge attached. It only weighs 1 pound 1 ounce. It's an ideal one-person stove, for spring and summer use. It costs under $20.

EVERYTHING DOESN'T HAVE TO BE NEW: BUYING USED EQUIPMENT

When people first get interested in backpacking, or even after they've been backpacking several years, they seem to have a fetish about buying new equipment. As far as we're concerned, that's utter nonsense. While it's pleasurable to buy something new every so often, there's absolutely no reason why you have to purchase everything brand new. Anna is particularly adept at buying second-hand equipment that's in excellent condition. She's paid less than half price for almost everything—from a super-deluxe Eddie Bauer 3-pound down bag ($50.00 for it last year), to a very handsome Alpine Designs daypack ($14 about three years ago).

Our point is that you can (and should) cash in on other people's mania for the newest and the latest in gear.

―――――――――――― **The Sources of Used Equipment**

YOUR FRIENDS AND RELATIVES

Most families have at least one person who was once an outdoor enthusiast. If you don't know offhand who that person is, ask your mother.

Your relatives may have military surplus gear or old goose-down hunting gear stored away somewhere. After a good cleaning, there's no reason why you shouldn't be able to use some or all of it.

Once you've exhausted your relatives as a source of supply, try your friends. Find out how old their equipment is, ask how much they paid for it, then sit down with an REI, Holubar, or EMS catalog and see how much it costs today. You should be able to buy it for less than half price.

RETAIL STORE ANNOUNCEMENTS

Read the bulletin board in your local backpacking equipment store, where used equipment is often advertised. Here's another method; we've found it works well: Advertise the piece of equipment you want to buy. Cut out a picture of it (look for one in a mail order catalog), paste it to a large index card, and write WANTED over it. Just write "looking for a tent" (or whatever it is you want) on the bottom of the paper, and make a fringe on the bottom of the paper with your telephone number written on each section of the fringe. If there is someone out there with a tent or other item to sell, they'll let you know.

COLLEGE AND SCHOOL BULLETIN BOARDS

These are also good sources of used equipment. Check them out even if you yourself don't attend that school. You can advertise "equipment wanted" here, too.

SUPERMARKET AND HEALTH FOOD STORE BULLETIN BOARDS

Used equipment may be advertised in supermarkets and health food stores. If you live in a large urban area, though, don't ask the seller to come to your house; go to his or her house to look at the equipment and take a friend with you. It may sound like paranoia, but that's urban reality.

CLUB NEWSLETTERS

If you don't already belong to an outing club, you should consider joining one. One of the benefits of membership in clubs like the Mountaineers, or the Appalachian Mountain Club, is free or low-cost classified advertising space in club publications. These are a *great*, very reliable place to advertise for second-hand equipment.

MAGAZINE CLASSIFIED SECTIONS

In our experience, it has been worthwhile to use the classified advertising sections of the following magazines when you want to buy or sell used equipment.

Backpacker Magazine
Classified Advertisements
65 Adams Street
Bedford Hills, New York 10507
($1.50 per word, $18.00 minimum)

Climbing Magazine
Classified Advertisements
Box E
Aspen, Colorado 81611
(30¢ per word, $3.00 minimum)

Summit Magazine
The Marketplace
P.O.Box 1889
Big Bear Lake, California 92315
25¢ per word

Wilderness Camping
Classified Advertising
1597 Union Street
Schenectady, New York 12309
65¢ per word

What to Look for When Examining Used Equipment

When buying used equipment the most obvious thing to look for is the general condition of the equipment—and some indication as to how it's been stored. If the owner of a secondhand down parka, for example, hands it to you tightly compressed into a stuff sack, chances are that it's been there for some time, and the loft isn't what it once was. If it's dirty and oily, it's obvious the person hasn't cared for the garment properly, and may have let dirt affect the loft too. But don't let a little dirt stop you from buying a good garment; it may in fact help bring down the price of the parka.

USED SOFT PACKS

Soft packs tend to be used and abused much more easily than any other piece of equipment. These are the things we recommend that you check first: the shoulder seam attachment joints. See if the pack has been repaired, and if the seams are intact. Look all along the inside seams of the pack, checking for rips and parting fabric. Inspect the lower shoulder strap contact points. Turn the pack upside down to check for any breaks or tears in the bottom fabric. Next, try all zippers and closure points. If the fabric is a waterproof material, run your nail over the inside of the fabric to see if there's any separation of the waterproof coating from the fabric itself. Decide what price you are willing to pay for the pack in A-1 condition. Take off 20 percent for each point mentioned above that does not meet with your approval. That way even if you do need a professional repair job, you won't have to pay more than the value of the soft pack for it.

Alpha/Special
(Photo courtesy of Hine/Snowbridge)

USED FRAME PACKS

Visually inspect the pack frame and pack for any obvious defects. Remove the pack from the frame. Check the pack for rips, or stitching weakness points. Do this inside and outside of the pack. Check all zippers to see that they're functioning correctly. If the pack has any leather patches, check to see that they're properly attached and not rotten. Check the grommets and attachment points of the pack to the frame for any undue wear where metal has touched nylon fabric. Aluminum oxide deposits on your pack are a potential weak point. A careful visual inspection will give you an indication of how much damage has been done.

Next, check the frame carefully. Has it been bent and straightened? If so the frame has been weakened. It is virtually impossible to straighten a bent frame without leaving an indentation. If the frame has been bent and straightened, cut offered price by 50 percent. If the seller wants to negotiate at that point, and if you feel it's worth another 10 percent, make a counter offer. But before you do, check the welds. If a weld has broken do not purchase the frame. If the frame is constructed with screws, buckles, and bolts, it wasn't any good to begin with; don't buy it. Examine the bottom of the frame at the contact points that touch the ground, since this area receives the most wear. If the tubing is crushed at the bottom, but the rest of the frame is in good condition you should still consider buying the frame pack—at a lower price.

Check the waistbands. Many would-be sellers of pack frames substitute cheaper waistbands from an older model. If that's been done reduce your offer by 20 percent. Later you can go out and buy a better waistband, which isn't that cheap.

Check all the clevis pins to see if they've worn larger holes into the frame. If greatly damaged, don't invest in the frame.

Reassemble the pack and frame; this will help you become familiar with them. Never make an offer until you know the seller's asking price. Point out all the defects you found in the pack and frame. You don't have to be abrasive when you do this; just be firm.

Knock 10 percent off the seller's asking price for each defect.

USED BOOTS

The idea of buying used boots may seem unappealing to you at first. But there is good reason to consider this purchase. You can get a well-broken-in pair of backpacking boots in good condition that are still serviceable—and you can get them for one-third the price of new boots. Try the boots on wearing two pairs of socks, one thin, one thick. Walk in them. If they don't feel comfortable, don't spend any more time with them. If they feel comfortable, check the tongue attachment to see that it is well anchored to the arch of the boot. Check the eyelets to see if they are intact. Put your hand all the way inside the boot to determine whether there are any rips or tears in the leather. Turn the boot upside down, and examine the tread on the Vibram sole; has it worn down to such a degree that it is no longer serviceable? Check the welt of the boot for any separation of mid-soles, leather, or the boot body. Check the boots for signs of deeply cracked leather—an indication that the boot may have been dried next to a fire.

If the damage is minimal, and the boots feel good on your feet, buy them. Once you get them home do the following: Wipe the inside down with a mild solution of alcohol; spray the boot with a foot fungicide; shake Dr. Scholl's foot powder into them; scrape off the residue of old waterproofing that may have collected in the seams; and saddlesoap the boots. Then completely waterproof your used boots. If minor repairs need to be made, consult the list of boot repair cobblers in Chapter 2.

USED TENTS

The basic rule when shopping for a used tent is always *set it up* before you buy. In rigging the tent, you'll have the opportunity to check all the pole connections, the elasticity of the shock cord, the stitching of the pole sleeves, the condition of all the side peg connections and guylines, the condition of the tent fabric, doors, walls, mosquito netting, and floor.

Before putting on the rainfly, get inside the tent. Zip the door and the mosquito netting closed. Examine the seams and the tent fabric: look for light coming through. Run your hands over the floor of the tent, checking for rips and tears. With a piece of tailor's chalk, make a circle around any damages. Come out of the tent, set up the rainfly, and inspect it.

If a tent is in good condition, it retains about 75 percent of its value. Find out the age of the tent and ask the seller if he is the first owner. If the answer is no, the value of the tent is only 45 percent of the cost of a new tent. Decide how much you want to pay for the tent, and how much the seller wants. Deduct 10 percent of the asking price for each major repair needed (zipper replacement, mosquito net replacement or repair, damaged tent fabric). If the seller agrees to your fair offer, don't hesitate.

USED SLEEPING BAGS

Down is valued almost like gold today. So if you find a used down-filled sleeping bag for sale, consider it seriously. Determine the brand of the bag and the year the original owner purchased it. You should inspect the bag and take note of the way it has been stored. If the seller pulls the bag out of a tight little stuff sack, chances are that it has been improperly cared for for some time. Remove it from the bag and lay it on a flat surface. Do not shake or fluff the bag. Looking at the unfluffed bag will give you an indication of the resiliency of the down. Now fluff the bag and see what kind of loft it really has.

A sleeping bag is like an article of clothing, a garment worn by the sleeping backpacker, so you will want it to be well cared for and clean. One way to tell whether the bag has been cleaned is to smell it—especially the foot section.

Turn the bag completely inside out, zip it up, and inspect the zipper and draft tube. Check all sewn-through stitching to see that it's intact. Ask the seller if the bag has been zippered onto another bag; chances are if it has, some of the baffle may be torn. While the bag is still inside out, pass the bag over your head under a naked lightbulb; you'll be able to see any thin spots.

Turn the bag right side out. Get in it and close it up around you. See how easy (or hard) it is to draw the zipper closed and tighten the hood.

If the bag is in reasonably good condition and you decide to make an offer for it, here's the best way to go about it. Know the cost of the bag when new and offer the owner half that price. You can haggle up from there. If the bag needs minor repairs, tell the seller that you want him to have the repairs made first. Most of the time the seller won't want to bother, and will be happy to lower the price. What we often do while we're at the seller's house is call the local repair shop for an estimate on the repairs that are needed. Then we tell the seller how much it'll cost to repair the bag. We deduct that amount right off the top. Then we go home and make the repairs ourselves.

This same advice applies to buying used synthetic-fill sleeping bags. However, it's very hard to judge the original loft unless you know what the bag looked like when new. Bear in mind that you can increase the loft of a Hollofil, Fiberfill, or Polarguard sleeping bag by adding a little fabric softener to the rinse water when you wash it.

USED PARKAS AND VESTS

Parkas and vests receive considerable abuse. They are easily damaged by sparks emanating from campfires, and cook stoves that flare up.

If the vest or jacket is presented to you in a stuff sack, beware. It has probably been there for weeks or months. Carefully remove the garment from the stuff sack and spread it out on a table without fluffing it. Note how quickly the garment regains its natural loft.

Open up the pockets and put your hands inside; dirt, cookie crumbs, or other garbage will indicate that that person hasn't really cared for the parka or vest. Tell the owner of the garment that it needs to be dry-cleaned. Deduct $6 to $10 from the asking price.

Turn the garment inside out. Look for cuts or tears and for evidence of any previous repairs. If a down garment has lost any of its filling take another 5 percent off the price. Hold the garment up to the light to check for any thin spots. If a major baffle is ripped and there is no down in the baffle, deduct 15 percent; down is very expensive, and so is labor. If it's a Polarguard or Hollofil garment, none of the fill should be missing; if it is, deduct 10 percent.

Put the garment on. Check to see that the zipper and all the snaps work. Be sure that all the drawstrings and head closures work properly. If a drawstring is missing, it's very easy to replace.

Check to see if the elastic on the cuffs is stretched out. Replacing the elastic is an annoying repair job; if it has to be done deduct 5 per cent per sleeve.

Point out all the damages you see to the owner of the item to make a case for the lowest price.

If you have the opportunity to buy a damaged down parka cheaply, do it. The down inside the garment is worth more than the shell, and it's good to have a stock of down on hand for repairs.

8

HOW TO FORM AN EQUIPMENT-BUYING COOPERATIVE

Over the past ten years we've seen people try all kinds of ploys to save money on their equipment purchases. One of the most exciting approaches we've seen is the incorporated cooperative buying association. Two of the earliest incorporated co-ops in this country were REI in Seattle, Washington, and Co-Op Wilderness Supply in Berkeley, California. Both of these very prosperous commercial outdoor suppliers started as small, localized groups looking for a viable new way to save money on their outdoor gear.

With some organizational skill, and a reasonable commitment of time, you and your friends can start a co-op too. There are two types. . . .

───────────How to Form a Nonincorporated Buying Association

In some states, it's illegal to use the word "cooperative" unless you're incorporated. So if you don't intend to incorporate, just to be on the safe side use the word "association" to describe what you're doing. To form a nonincorporated buying association, all you have to do is get together with a group of about six friends and chip in $25 each. Take your $150 and open a checking account in the name of the association at a local bank. Be sure that this is *not* a business account; it should be a personal checking account with the name of the authorized buyer printed above the name of the association. Use your authorized buyer's home address as your association's address.

Next, order a limited quantity of stationery and envelopes printed with the name and address of the association. Now you're ready to start operating. Write on your association letterhead to the various manufacturers of equipment you want to buy in quantity.[*] If you are not interested in buying more than one or two of an item, don't even bother contacting the company.

Many manufacturers will not want to deal with you directly until one of their salespeople has spoken with you (you'll want to avoid the salesmen; they tend not to like to deal with co-op buying groups).

Once you've obtained wholesale price lists and order blanks, you must get the people in your association to *pre-pay* their orders. But first buy a rubber stamp printed with the name and address of your association, an inkpad, and about 50 numbered purchase orders at a local stationery store. Stamp the purchase orders with the rubber stamp. Type the orders out on these purchase order forms and send them, accompanied by a letter (on association stationery) from your authorized buyer, and a check, to the manufacturer. Be sure that association members understand that the wholesale price printed in manufacturers catalogs is *not* the price at which you'll be buying; added to the wholesale price is an extra 10 percent to cover postage and shipping cost.

Caution

You should *not* resell equipment you buy in this way to anyone outside your association, and especially not to anyone at a profit. If you do, *you* (the purchasing agent) are liable for damages should anyone be injured while using any of the equipment. You also become legally responsible for filing income tax returns showing your profit.

A nonincorporated buying association is a very informal organization that may tend to break down once it has made the initial equipment purchases. Association members will also have to deal with the personal equipment fetishes and inevitable hunger for power of the purchasing agent. There's no way to prevent this, except to form an incorporated buying cooperative.

———————— The Incorporated Buying Cooperative

There are several reasons why an incorporated buying cooperative is superior to a nonincorporated buying association. First, it is run by its members, and no one individual can take control. This also means that if the purchasing agent begins to lose interest in the

---

[*] *The Chambers Sourcebook* contains a complete list of equipment manufacturers and their addresses, many of which appear in this book, too. (See p. 116)

organization, your co-op can still survive. It means too that all the purchases are in the name of the organization—not in the name of one person. The advantage of this is that even if the leadership changes, the co-op won't have to renegotiate for the right to purchase directly from certain manufacturers who are used to dealing with one particular person. It also protects the individual members because in an incorporated buying cooperative you have bylaws, state procedures, elected officials, and you have annual meetings and reports. An incorporated buying cooperative can be a very democratic organization if the amount of time and energy each person has to put in is clearly defined right from the start. Incorporated buying cooperatives can also exchange information with other co-ops across the country, and they are entitled to membership in the Co-op League of the United States, located at 1828 L Street, N.W., Washington, D.C. 20036.

Forming an incorporated buying cooperative is inexpensive. The major cost involved will be legal fees for drawing up the incorporation papers. You should be able to incorporate for as little as $125 (less if you bring an attorney into your co-op). Follow the lawyer's advice when drafting the bylaws, but suggest to your attorney that you file your incorporation papers in Washington, D.C. It's easier and cheaper to be incorporated there, and the incorporation is valid in every state of the union.

After you incorporate, open a business checking account in the name of your incorporated buying cooperative. You should require two authorized signatures on every check. The only *printed* name on these checks should be the name of your co-op, not the names of individual purchasing agents. We suggest that you rent a post office box for a year at your local post office and notify U.P.S. of a delivery address for any packages that will be shipped to your co-op; this last step is particularly important because U.P.S. will not deliver to a post office box.

Your co-op will need printed stationery, purchase orders, and a rubber stamp with the name of the co-op. The amount of membership dues will depend upon the number of members, and upon operating costs.

If your co-op has a large membership, it means that certain people within the co-op will have a lot of responsibility and work to do; your co-op may decide to pay those people for their work, or to give them a credit toward equipment purchases.

Other Sources of Information on Forming Equipment-Buying Cooperatives

Check with your local librarian for any information he or she may have on forming co-ops. Check the *Readers' Guide to Periodical Literature* for listings under cooperatives. In addition, you'll probably want to send a request for information accompanied by a stamped self-addressed envelope to the following:

Co-op
c/o Silsu-Olana
Route 45
Pomona, New York 10970

Consumer Co-op Alliance
Co-op Services, Inc.
7404 Woodward Avenue
Detroit, Michigan

The Co-op League of
the U.S.A.
1828 L Street N.W.
Washington, D.C. 20036

The Department of
Agriculture and Markets
State Capitol
Albany, New York

Johen Klein
North American Student Co-op Association
530 South State Street
Ann Arbor, Michigan 43109

9

BLUE-CHIP OUTDOOR GEAR / WHAT TO BUY NOW AND PUT AWAY FOR LATER

The cost of much outdoor gear has been going up at a rate that exceeds our current rate of inflation. Down products especially, as we previously mentioned, have been appreciating in value at a very rapid rate.

What follows is a list of outdoor gear which you should buy now at sales and in bargain basements, even if you don't need it. If you have any extra spending money, it really pays to invest in these items *now*—and put them away for future use. If need be, you can always sell them to your friends.

All Down Items

If you have any fondness for down, now is the time to buy it. Barring a sudden population explosion of geese and ducks, down prices will never be lower than they are now (not that they're so cheap now). We recommend that you search thrift shops for old down parkas; remove the filling from these items and store it in breathable large stuff sacks. Eventually, you'll be able to use it to repair the down garments you now own, or to make a nice down overbag.

Wool Sweaters

Dachstein sweaters are especially valued. These sturdy winter mountaineering sweaters have gone up over $20 in price over the past two years. You can never have too many sweaters, especially if it is as fine a garment as the one made by Dachstein.

Wool Shirts

With the growth of cross country skiing, winter backpacking and camping, and other winter sports, there has been an increased demand for durable, warm woolen shirts. During the winter of 1978 the price of Woolrich shirts went up 15 percent. Look for sales, especially those that feature the Woolrich Alaskan and Buffalo shirts. These are warm, handsome garments that will last you many years.

Wool Socks

This may seem at first like a silly suggestion. But those beautiful Norwegian Ragg socks are getting prohibitively expensive. As we write this, they cost as much as $9.00 a pair! If you can, it pays to put a few pair away in your closet; be sure to renew the camphor balls annually. Other wool socks are not quite as expensive—but a good pair of wool knicker socks is not cheap. We've never seen the Norwegian Ragg socks on sale, but Anna has found many pairs of fine knicker socks in clearance bins and at clearance sales in Vermont and New Hampshire. It'd be worthwhile for you to pick up several pair of these and salt them away too.

Dachstein Mitts

Dachstein mittens and gloves are another item that seldom appears at sales, so you may as well buy now and put them away for future winters. They're appreciating in value at more than 10 percent per year.

Craghopper Knickers

The best woolen knickers we've worn are the Clark's Craghopper Knickers. They are the most popular knickers in New England—and they've increased in price from $24 to $38 a pair between 1977 and 1978. They're also becoming scarce.

Supergaiters

If you can find any Chouinard Supergaiters, grab them. They've been discontinued (believe it or not).

Molitor Boots

These are still the best ice-climbing boots you can buy. They're no longer being imported into the United States, so chances are they will begin to turn up in clearance centers and bargain basements.

EB Climbing Shoes

These have been going up in price every year. The extra pairs Anna bought on sale for $23 each three years ago are now worth close to $45. Extra EB's are a good investment if you climb.

Royal Robbins Yosemite Boot

These are now out of production, but you can still occasionally find a few pairs in the clearance centers and bargain basements listed in the next chapter. These were once considered the most versatile combination climbing and lightweight hiking shoe available in the United States. At one time, if you had a pair of Robbins, you were "in the know." These are still good quality boots—and an especially good buy if you find them marked down in a clearance center.

Stoves

If you find Optimus 8R or Optimus 111B stoves on sale now, buy a couple. You won't be sorry. Within the next five years, the price of these stoves is sure to increase about 20 percent. Look for spare parts, too; these are common in bargain basements.

Fuel Bottles

Sigg fuel bottles have gone up 200 percent in price over the past three years. Unused fuel bottles make excellent canteens.

Sigg Cookkits

All imported Swiss or Austrian cookkits have been soaring in price. Their quality is exceptional. These make very good investment gear.

Poly Bottles

Imported poly bottles, especially the Austrian wide-mouth bottles, are being pushed out of the market by American-made copies. The quality of the American-made products is, in this case, not as high as the Austrian: The walls are not as thick, and they don't last as long.

Optimus 8R
(Photo courtesy of Optimus International)

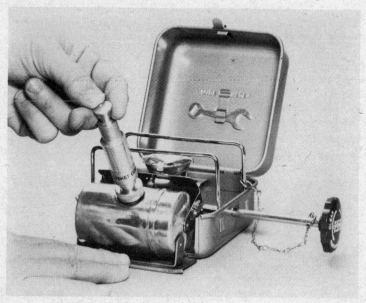

Tents

Tents have increased in price every year for at least the past six years. The prices of some tents, like the North Face St. Elias tent, have doubled over the past five years! Don't ignore the opportunity to purchase an extra tent—either secondhand or new.

Nylon Stuff Sacks

Nylon stuff sacks are one of the most useful backpacking and camping items one can purchase.

Karrimor Joe Brown Backpacks

This is one of the most versatile all-purpose hiking, backpacking, and climbing packs available in this country. They're generally in short supply. Over the past three years, the cost of this pack has gone up over $30.

Wooden Cross-Country Skis

It's getting harder and harder to find wooden cross-country skis. Steve's first pair of wooden touring skis cost less than $5 ten years ago in a Stowe, Vermont, thrift shop. Today, similar quality skis cost about $80 new. These wooden touring skis are beautiful; they're more aesthetically pleasing than fiberglass skis, and they work just as well. You might try searching thrift shops for them.

Elite Bernina Rope
(Photo courtesy of
Kalmar Trading Corporation)

Alpine Eyeglasses

Alpine eyeglasses made with ground glass lenses have virtually disappeared from the shelves. They've been replaced by cheap plastic lenses that scratch easily, and frames that self-destruct. You can still find some Alpine eyeglasses with glass lenses—for less than $20.

Ice Axes

A good reliable ice ax is essential for winter backpackers. Ice axes keep going up in price—about a 50-percent increase over the past four years. If you have a favorite wooden shafted ice ax, especially a Chouinard or an Interalp ax, now is the time to buy an extra one for the years ahead.

Climbing Ropes

It wasn't so long ago that we were paying $35 for a 150-foot llmm Edelrid climbing rope. They now cost close to $100. If you climb you'll see the sense of buying an extra one—especially if it is on sale.

10

STATE-BY-STATE GUIDE to SALES, DISCOUNTS, MANUFACTURERS' OVERRUNS AND SECONDS, CLEARANCE CENTERS, REPAIR SERVICES, AND RENTAL DEPARTMENTS

What follows is a detailed consumer's guide for backpackers. This is the first time this information has ever been available.

It tells you which retailers have sales and when—so you can plan to buy your equipment while it is on sale.

It tells you where the stores are that offer backpacking gear at less than manufacturers' suggested list price. It tells you how much of a discount they offer and (as often as possible) the specific brands you'll find there.

If you're shopping for a nonprofit organization, we list stores that give special discounts to religious groups, schools, and outing clubs. At a glance, you'll be able to determine who near you gives the best discount.

We list hundreds of stores that carry manufacturers' overruns and seconds at considerable discounts. Not all of the stores listed will have this type of merchandise at all times—but at least now you'll know where to look for it.

Also, we list many bargain basements and clearance centers. These are often treasure troves for backpacking gear bargain hunters. Some, like Eddie Bauer's Seattle outlet, The Loft, and Eastern Mountain Sports' Ardsley, New York, and Boston Bargain Basements are worth a trip of many miles. They're consistently ripe with bargains.

We list repair services across the country because one of the best ways of saving money on backpacking equipment is to maintain the gear you already own. These are also essential services if you are wise enough to buy some of your gear secondhand.

This chapter also covers hundreds of rental services across the United States. Make good use of them. There's no reason to lay out good money for items like crampons, ice axes, or even cross-country skis, if you can rent them easily and inexpensively the three or four times a winter you may use them. Renting items like tents, sleeping bags, and even backpacks is a great idea if you're a beginning backpacker; you won't have to rush into major purchases before you're absolutely sure what you want to buy—or before the stores near you have their regularly scheduled sales.

Some stores have "Rental Sales" listings. This means that the store will sell off its used rental gear. This is an *excellent* way to buy cross-country skis, stoves, ice axes, and tents at what is usually a *very* good price.

In the guide that follows, you will frequently see the initials *R.P.* after a store listing. These letters indicate that that store will apply all or part of your rental fee toward the purchase of new equipment, or used rental equipment. Please note that *R.P.* usually does not apply to times of sales or clearances.

You will find this guide particularly helpful if you are traveling cross-country. Both of us enjoy stopping by backpacking shops that are new to us; they're a great source of local information—and we very often find bargains.

To compile the information that follows, we contacted over 2300 backpacking and camping equipment retail stores. Over 600 stores were contacted twice. If a store near you does not appear in the guide that follows, chances are they just didn't bother to respond to our questionnaire. Should you know of any stores not listed here that offer substantial discounts, please do let us know; we'd be glad to hear about them. You can write to us at: Box 261, Franconia, New Hampshire 03580.

We'd like to explain a bit more about manufacturers' overruns. Many large retail outlets and camping chain stores have equipment produced for them by large manufacturers. If an order for 5,000 sleeping bags is placed with the manufacturer, it's not uncommon for the manufacturer to produce 400 or 500 excess bags with the same specifications to allow for any damage or soiling. Store jobbers approach these manufacturing companies and purchase these overruns. In effect, what you get is a first-quality bag at a savings of up to 40 percent. This is true of everything from tents to stoves to sleeping bags and boots.

Manufacturers' seconds are another good buy. In the manufacturing process, nylon can be stained or soiled pretty easily. In the case of tents, such manufacturers as Jansport frequently sell their blemished tents to the chain stores, who can then frequently sell them at up to a 50 percent discount. In garments, seams may be sewn crookedly. This may not affect the fit and function of the gear, but it still can't be sold as a first-quality item. For 30 to 50 percent off, we don't mind if the seams of our down vests are an eighth of an inch off.

In boot manufacturing, particularly, it's not that uncommon for the leather to become scraped. These boots are usually not available through your retailer, but they often appear at factory outlet stores at a much lower price. You're going to scrape your boots anyway, so what difference does it make if you buy them already pre-scraped? Don't hesitate to purchase seconds; the really badly damaged merchandise won't turn up in your retail store.

Nothing beats careful sale shopping. But there are certain rules you must observe. First, know what you want, and know how much it costs at regular retail price. An informed shopper makes the most intelligent purchasing decisions and can save the most money. Read as many mail order catalogs as you can. Don't be embarrassed about going to your local backpacking store and trying on several different parkas without buying one. Next, you *must* wait for the sale. Don't succumb to paying full retail price. Find out exactly when the sale will be held. We list many dates in the chapter that follows —but dates may change so ask the store salespeople when the sale will be held. Before the sale begins, narrow your choice to two items, two tents for instance or two parkas. Also, familiarize yourself with the layout of the store so that when the sale begins, you know exactly where to go. Arrive *several hours* before the store opens on sale day. We once spent eighteen hours in front of Eastern Mountain Sports in Boston; it was fanatical, but fun. But you don't have to go to such extremes. Just make sure that you are among the very first on line. When the doors open, go right to the area where the gear you want is being sold. Grab two, just in case one is damaged. Make a rapid inspection, pick one, pay, and *leave the store.* That way you won't buy things you don't need.

————————————————————————— **Alabama**

BIRMINGHAM

ALABAMA OUTDOORS   205-591-7067
115 Century Plaza
Birmingham, Alabama 35211
   Discount to Nonprofit organizations: 10 percent. *Rent:* Sleeping bags, stoves, tents, frame packs.

BLUE WATER OUTDOOR EQUIPMENT, INC.   205-870-1919
527 Brookwood Village
Homewood (Birmingham), Alabama 35209
   Repairs: Will arrange for equipment repairs. *Rent:* Sleeping bags,
foam pads, frame packs, tents. *Rental Sale:* Each spring. *R.P.*

MOUNTAIN BROOK

BAIRS SKI AND HIKE SHOP   205-871-2247
2418 Canterbury Road
Mountain Brook, Alabama 35223
   Manufacturers' Seconds: 25 percent off list price. *Rent:* Downhill skis, frame packs, soft packs, tents. *R.P.*

--- **Alaska**

ANCHORAGE

ALASKA MOUNTAINEERING AND HIKING   907-272-1811
2633 Spenard Road
Anchorage, Alaska 99503
   Repairs: Cross-country skis. *Rent:* Cross-country skis, ice axes,
tents. *Rental Sale:* Each spring. *R.P.*

MUSKEG OUTFITTERS   907-344-8438
10197 Klatt Station
Anchorage, Alaska 99502
   Discount to Nonprofit Organizations: 10 percent. *Rent:* Tents,
stoves, snowshoes, frame packs, ice axes, crampons, cross-country
skis.

TWO WHEEL TAXI & SKI SHOP   907-272-4431
405 East Northern Lights Boulevard   907-278-4578
Anchorage, Alaska 99503
   Semiannual Sales: Fall and spring. *Discounts:* Some merchandise
is sold for less than manufacturer's suggested retail price. *Manufacturers' Seconds and Overruns:* 20 percent to 50 percent off list
price. *Repairs:* Skis, ski boots, bindings. *Rent:* Sleeping bags, tents,

frame packs, soft packs, snowshoes, downhill skis, cross-country skis.

—————————————————————————————**Arizona**

FLAGSTAFF

THE ALPINEER   602-774-7809
406 South Beaver Street
Flagstaff, Arizona 86001
   Discount to Nonprofit Organizations: Only on purchases over $500. *Manufacturers' Overruns and Seconds:* Occasionally sold at a 20 percent discount. *Repairs:* Minor repairs. *Rent:* Snowshoes, ice axes, cross-country skis. *Rental Sale:* Spring. *R.P.*

GLENDALE

YATES STORES   602-939-6531
6048 North 59th Avenue
Glendale, Arizona 85301
   Discounts: 20 percent to 30 percent discount on name brand merchandise. *Discount to Nonprofit Organizations:* Additional 10 percent discount. *Manufacturers' Overruns and Seconds:* 40 to 50 percent less than manufacturers' suggested list price. *Clearance Center:* Reductions of from 40 to 50 percent on brands like Coleman, Camp Trails, White Stag. *Repairs:* Minor repairs. *Special Features:* This store offers some of the most substantial equipment savings in the United States: an excellent place to stop while traveling.

PHOENIX

DESERT MOUNTAIN SPORTS   602-265-4401
4506 North 16 Street
Phoenix, Arizona 85016
   Annual Sale: They don't have a regular schedule of sales. Check with the store. *Discount to Nonprofit Organizations:* 10 to 20 percent discount. *Rent:* Frame packs. *Rental Sale:* They occasionally sell their rental equipment. Ask them.

HOLUBAR MOUNTAINEERING   602-955-3391
3925 East Indian School Road
Phoenix, Arizona 85018

Semiannual Sales: October and March. *Discounts:* Their own brand of merchandise regularly represents a 10 to 20 percent saving over comparable brand name merchandise. *Discount to Nonprofit Organizations:* Varies from organization to organization, and is at least partially dependent upon the size of the order. *Manufacturers' Overruns:* They sell manufacturers' overruns at about 20 percent off list price—including their own brand. *Clearance Center:* Their clearance area can contain brands like Fabiano, Pivetta, Kelty, and Wilderness Experience—at 20 to 30 percent off list price. *Repairs:* They repair their own brand of equipment. Also stoves and skis. *Rent:* Sleeping bags, foam pads, tents, frame packs, soft packs, ice axes, cross-country skis, snowshoes. *Rental Sales:* October and March. *R.P.*

SKI HAUS ACTION SPORTS   602-955-4081
2501 East Indian School Road
Phoenix, Arizona 85016

Semiannual Sales: Spring and fall. *Discount for Nonprofit Organizations:* Varies between 10 and 20 percent. *Manufacturers' Overruns and Seconds:* They do carry manufacturers' overruns and seconds but didn't tell us how much of a discount they offer on this merchandise. *Clearance Center: Very* good discounts, about 50 percent off list price. Merchandise may include such brands as Raichle, Jansport, North Face, Blacks, and Nordica. *Rent:* Tents, frame packs, downhill skis, cross-country skis. *Rental Sales:* End of each season. *R.P.*

YATES STORES   602-955-3360
3931 East Thomas
Phoenix, Arizona 85040

Discounts: 20 to 30 percent discount on name brand merchandise. *Discount for Nonprofit Organizations:* 10 percent. *Manufacturers' Overruns and Seconds:* 40 to 50 percent less than regular price. *Clearance Center:* 40 to 50 percent off list price *Repairs:* Minor repairs.

YATES STORES   602-277-5415
4750 North 16th Street
Phoenix, Arizona 85016

Discounts: 20 to 30 percent discount on name brand merchandise. *Discount for Nonprofit Organizations:* 10 percent. *Manufacturers' Overruns and Seconds:* 40 to 50 percent less than regular price. *Clearance Center:* 40 to 50 percent off. *Repairs:* Minor repairs.

TEMPE

THE GROVER COMPANY   602-967-8738
2111 South Industrial Park Avenue
Tempe, Arizona 85282
   Discounts: 15 percent below manufacturers' suggested retail price on Mountain House Freeze-Dried Foods. *Discount for Nonprofit Organizations:* An additional 10 percent off.

HOLUBAR   602-968-2712
1043 East Lemon
Tempe, Arizona 85282
   Semiannual Sales: October and March. *Discounts:* Their own brand of merchandise regularly represents a 10 to 20 percent saving over comparable brand name merchandise. *Discount to Nonprofit Organizations:* Varies from organization to organization, and is at least partially dependent upon the size of the order. *Manufacturers' Overruns:* They sell manufacturers' overruns at about 20 percent off list price—including their own brand. *Clearance Center:* Their clearance area can contain brands like Fabiano, Kelty, and Wilderness Experience—at 20 to 30 percent off list price. *Repairs:* They repair their own brand of equipment, also stoves and skis. *Rent:* Sleeping bags, foam pads, tents, frame packs, soft packs, snowshoes, ice axes, cross-country skis. *Rental Sale:* October and March. *R.P.*

SKI HAUS ACTION SPORTS   602-967-7442
705 South Forest Avenue
Tempe, Arizona 85281
   Sales: Spring and fall—call them to check exact time. *Discount to Nonprofit Organizations:* 10 to 20 percent. *Manufacturers' Overruns and Seconds:* They carry both, but at an unspecified discount. *Clearance Center:* Up to 50 percent savings on brands like Kelty, North Face, Jansport. *Rent:* Tents, frame packs, downhill skis, cross-country skis. *Rental Sales:* End of seasons. *R.P.*

TUCSON

SKI HAUS ACTION SPORTS   602-881-4544
2823 East Speedway
Tucson, Arizona 85716
   Semiannual Sales: Spring and fall—call them to check the exact time. *Discount to Nonprofit Organizations:* 10 to 20 percent. *Manufacturers' Overruns and Seconds:* They carry both, but at an unspecified discount. *Clearance Center:* Up to 50 percent savings on brands like Kelty, North Face, Jansport. *Rent:* Tents, frame packs, downhill skis, cross-country skis.

--- **Arkansas**

FAYETTEVILLE

OZARK MOUNTAIN SPORTS   501-521-5820
226 North School
Fayetteville, Arkansas 72707
   Manufacturers' Overruns and Seconds: 20 percent off list price. *Rent:* Sleeping bags, tents, frame packs, kayaks.

THE PACKRAT   501-521-6340
2950 North College
Fayetteville, Arkansas 72701
   Discounts: 10 to 15 percent off manufacturers' suggested list price on some items. *Discount to Nonprofit Organizations:* 10 percent for church groups. *Repairs:* Canoes and kayaks. *Rent:* Tents, kayaks, canoes. *Rental Sale:* End of each summer. *R.P.*

FORT SMITH

OZARK MOUNTAIN SPORTS   501-783-6667
4117 Grand
Fort Smith, Arkansas 72901
   Manufacturers' Overruns and Seconds: 20 percent off list price. *Rent:* Sleeping bags, tents, frame packs, kayaks.

ANAHEIM

THE GRANT BOYS   714-821-8900
2636 West La Palma
Anaheim, California 92801
   Semiannual Sales: Washington's Birthday and Halloween. *Discount:* 15 percent off manufacturers' suggested retail, but on a limited range of gear. *Discount to Nonprofit Organizations:* 10 percent. *Repairs:* Coleman appliances. *Rent:* Sleeping bags, tents, stoves, frame packs, lanterns, Coleman lanterns, cots. *Rental Sale:* End of summer season.

ARCATA

ARCATA TRANSIT AUTHORITY   707-822-2204
650 10th Street
Arcata, California 95521
   Semiannual Sales: Spring and fall. *Discounts:* They do sell merchandise at less than manufacturers' suggested retail price; the discount varies with the merchandise. *Manufacturers' Seconds:* 20 to 30 percent off suggested list price. *Repairs:* Down equipment, bicycles, packs, camping equipment. *Rent:* Sleeping bags, foam pads, tents, stoves, frame packs, soft packs, snowshoes, canoes, ice axes, crampons, cross-country skis, bicycle packs, Gore-tex® shelters (one-person bivouac shelters). *Rental Sales:* End of each season. *R.P.*

BAKERSFIELD

BIGFOOT MOUNTAINEERING   805-323-9481
2594 Brundage Lane
Bakersfield, California 93304
   Rent: Sleeping bags, foam pads, parkas, tents, stoves, frame packs, soft packs, snowshoes, ice axes, cross-country skis. *Rental Sales:* End of each season.

BERKELEY

CO-OP WILDERNESS SUPPLY   415-843-9300
1607 Shattuck Avenue
Berkeley, California 94702
  Rent: Frame packs, sleeping bags, tents, stoves.

GRANITE STAIRWAY MOUNTAINEERING   415-848-7866
2160 University
Berkeley, California 94704
  Semiannual Sales: April and October. *Discount to Nonprofit Organizations:* 10 percent. *Manufacturers' Overruns and Seconds:* 20 to 40 percent off list price. *Clearance Center:* 20 to 40 percent off list price on brands like Fabiano, Wilderness Experience, Snow Lion, Class 5. *Repairs:* Skis, packs, bags, and all other repairs. *Rent:* Sleeping bags, parkas, tents, stoves, frame packs, soft packs, snowshoes, downhill skis, ice axes, crampons, cross-country skis, ski racks. *Rental Sales:* Only damaged equipment. *R.P.*

HARBERTS BROTHERS SPORTING GOODS   415-843-0717
2338 Shattuck Avenue
Berkeley, California 94704
  Annual Sale: March. *Manufacturers' Seconds:* 33 percent discount—one of the largest discounts in the United States on this type of merchandise. *Rent:* Sleeping bags, parkas, tents, stoves, frame packs, snowshoes, downhill and cross-country skis, sleds, ski racks, ski pants. *Rental Sales:* October and August.

MOUNTAIN TRADERS   415-845-8600
2140 Center Street
Berkeley, California 94709
  Manufacturers' Overruns and Seconds: 10 to 40 percent discount. *Clearance Center:* They do have a clearance area but didn't tell us what kind of reductions you can find there. *Rent:* Boots, sleeping bags, tents, frame packs, soft packs, snowshoes, ice axes, crampons, cross-country skis. *R.P. Special Features:* They'll sell your used equipment on a consignment basis.

NORTH FACE   415-548-1371
2804 Telegraph Avenue
Berkeley, California 94705
  Semiannual Sales: April and September. *Manufacturers' Sec-*

onds: 20 percent off list price—this is an excellent place to look for North Face's own brand. *Clearance Center:* 20 percent discount. *Repairs:* Cross-country skis. *Rent:* Sleeping bags, foam pads, tents, stoves, frame packs, soft packs, snowshoes, cross-country skis. *R.P.*

RECREATIONAL EQUIPMENT, INC.   415-527-4140
1338 San Pablo
Berkeley, California 94702
   Discount: Their own label equipment averages 20 percent below prices for comparable merchandise. *Repairs:* Skis, bicycles, stoves. *Rent:* Frame packs, soft packs, cross-country skis. *Rental Sales:* Two annual sales, spring and fall—ask them for this year's dates. *R.P. Special Features:* Recreational Equipment, Inc. is a co-op, so they pay a 10 percent dividend on purchases made by co-op members at the end of the year. This dividend can be applied toward equipment purchases or can be received in cash.

SIERRA DESIGNS   415-843-2010
4th and Addison Street
Berkeley, California 94710
   Semiannual Sales: Spring and fall, markdowns to 40 percent. *Discount:* Some merchandise is sold at less than manufacturers' suggested list price. *Manufacturers' Overruns and Seconds:* Sierra Designs equipment at 25 percent less than usual price—a *great* buy. *Repairs:* They repair their own line of outdoor gear. *Rent:* Sleeping bags, foam pads, tents, frame packs, soft packs, snowshoes, ice axes, cross-country skis, parkas, pots and pans. *Rental Sales:* End of each season. *R.P. Special Features:* They sell remnants of material for the do-it-yourselfer.

THE SKI HUT   415-843-6505
1615 University Avenue
Berkeley, California 94703
   Annual Sale: Early September. *Manufacturers' Overruns:* They have them occasionally. *Repairs:* Any repairs that do not require sewing. *Rent:* Down and synthetic fill sleeping bags, tents, stoves, frame packs, soft packs, snowshoes, downhill skis, ice axes, crampons, cross-country skis. *Rental Sale:* Early September.

ZIMM'S SURPLUS   415-549-3113
2136 University Avenue
Berkeley, California 94701

Discounts: 10 to 25 percent discounts on brands like Coleman, Twin Peaks, Slumberjack. *Discount to Nonprofit Organizations:* 10 to 20 percent. *Manufacturers' Overruns and Seconds:* 20 to 40 percent discount. *Clearance Center:* 20 to 50 percent off manufacturer's list price.

BEVERLY HILLS

BIKECOLOGY BIKE & BACKPACKER SHOP   213-278-0915
9101 Olympic Boulevard
Beverly Hills, California 90212
   Discounts: 15 percent discount on some brands. *Repairs:* All bicycle repairs. *Special Features:* They have an interesting "Anti-Rip-off Policy": If a bicycle is stolen during the 1st year, they'll sell you a replacement bicycle for their cost plus 10 percent.

SUNLAND SPORTS INC.   213-652-4990
8677 Wilshire Boulevard
Beverly Hills, California 90211
   Semiannual Sales: May, September. *Rent:* Sleeping bags, tents, frame packs, soft packs, downhill and cross-country skis, ice axes, crampons, skin diving equipment. *R.P. Special Features:* They make custom wetsuits.

BISHOP

BIKE AND SPORT SHOPPE   714-873-5070
772 North Main Street
Bishop, California 93514
   Repairs: Hang gliders, bicycles, cross-country skis. *Rent:* Sleeping bags, foam pads, tents, stoves, frame packs, soft packs, snowshoes, ice axes, bicycles, crampons, cross-country skis. *R.P.*

CAMPBELL

THE NORTH FACE   408-374-5205
349 East Campbell Avenue
Campbell, California 95008
   Semiannual Sales: April, September. *Manufacturers' Seconds:* 20 percent discount—look here for seconds of their own

fine brand. *Clearance Center:* 20 percent off list price. *Rent:* Sleeping bags, foam pads, tents, stoves, frame packs, soft packs, snowshoes, ice axes. *Rental Sales:* April and September. *R.P.*

CASTRO VALLEY

CO-OP WILDERNESS SUPPLY   415-886-4550
3667 Castro Valley Boulevard
Castro Valley, California 94546
   Annual Sale: July. *Discount:* 10 to 15 percent off manufacturers' suggested retail price on brands like Raichle, Mountain-10, Jansport, Twin Peaks, and Snow Lion. *Discount to Nonprofit Organizations:* 10 percent discount on food orders. 5 percent discount on equipment orders of $300 or more, 10 percent discount on equipment orders of $1000 or more. *Manufacturers' Seconds:* 20 to 70 percent off list price *Repairs:* They repair down clothing and sleeping bags. They also do cleaning of down garments and sleeping bags. Their cleaning service is very reliable. *Rent:* Parkas, frame packs, sleeping bags, tents, soft packs, cross-country skis, snowshoes, gaiters, ski racks. *Rental Sales:* October and May. *R.P.*

CERRITOS

NEAL'S SPORTING GOODS   714-924-1625
60 Cerritos Mall
Cerritos, California 90701
   Annual Sale: September. *Rent:* Sleeping bags, tents, frame packs, stoves, downhill skis. *R.P.:* One-half applied.

CHULA VISTA

STANLEY ANDREWS SPORTS OUTFITTERS   714-422-0662
485 H Street
Chula Vista, California 92010
   Discount to Nonprofit Organizations: 10 to 20 percent. *Rent:* Sleeping bags, foam pads, tents, stoves, frame packs, snowshoes, downhill and cross-country skis. *Rental Sales:* Periodically—call them. *R.P.*

COSTA MESA

THE GRANT BOYS   714-645-3400
1750 Newport Boulevard
Costa Mesa, California 92627

Semiannual Sales: Washington's Birthday and Halloween. *Discount:* 15 percent off manufacturers' suggested list price, but they carry a limited range of gear. *Discount to Nonprofit Organizations:* 10 percent. *Rent:* Sleeping bags, tents, stoves, frame packs, lanterns, Coleman lanterns, cots. *Rental Sales:* End of summer season.

COVINA

SANDY'S SKI RENTALS   213-332-4949
588 N. Azusa Avenue
Covina, California 91722

Rent: Sleeping bags, foam pads, parkas, tents, stoves, frame packs, soft packs, downhill skis, inflatable boats. *Rental Sales:* End of each season. *R.P.*

CULVER CITY

BIKECOLOGY BIKE & BACKPACKER SHOPS   213-559-8800
5179 Overland Avenue
Culver City, California 90230

Discount: 15 percent below manufacturers' suggested retail price on a few good brands including Snow Lion. *Special Features:* They have an interesting "Anti-Rip-off Policy": If a bicycle is stolen during the first year, they'll sell you a replacement bicycle for their cost plus 10 percent. It's a good thing they don't have a branch in New York City; they'd spend three days each week just replacing stolen bicycles.

CUPERTINO

ANTELOPE CAMPING EQUIPMENT   408-253-1913
21740 Granada Avenue
Cupertino, California 95014

Discounts: 10 percent discount on cash sales over $200, 15 percent discount on cash sales over $300. *Rent:* Frame packs, soft packs, snowshoes.

EL CAJON

STANLEY ANDREWS SPORTS OUTFITTERS   714-444-9346
478 Fletcher Parkway
El Cajon, California 92020
  Discount to Nonprofit Organizations: 10 to 20 percent.
Rent: Sleeping bags, foam pads, tents, stoves, frame packs, snowshoes, downhill and cross-country skis. *Rental Sales:* Periodically—ask them. *R.P.*

ESCONDIDO

STANLEY ANDREWS SPORTS OUTFITTERS   714-566-7010
360 West Washington
Escondido, California 92025
  Discount to Nonprofit Organizations: 10 to 20 percent.
Rent: Sleeping bags, foam pads, tents, stoves, frame packs, snowshoes, downhill and cross-country skis. *Rental Sales:* No set time—ask. *R.P.*

FULLERTON

FULLERTON INC.   714-992-2626
717 S. Harbor Boulevard
Fullerton, California 92632
  Rent: Cross-country skis, sleeping bags, foam pads, parkas, tents, stoves, frame packs, soft packs, downhill skis, inflatable boats. *Rental Sales:* End of each season. *R.P.*

GLENDALE

KELTY MOUNTAINEERING AND
BACKPACKING   213-247-3110
1801 Victory Boulevard
Glendale, California 91208
  Semiannual Sales: April and October. *Repairs:* Only Kelty products. *Rent:* Sleeping bags, foam pads, tents, stoves, frame packs, soft packs, ice axes, cross-country skis. *Rental Sales:* April and October. *R.P.*

VALUE CENTERS   213-243-7265
1133 West Glenoaks Boulevard
Glendale, California 91202
   Discount: 40 percent off suggested list price, but on a limited
range of merchandise. They do have Camp Ways, Coleman, and
Wenzel products. *Manufacturers' Overruns:* 50 percent off list
price.

GOLETA

UPPER LIMITS   805-967-0746
5783 Calle Real
Goleta, California 93017
   Rent: Sleeping bags, foam pads, tents, stoves, frame packs, soft
packs, snowshoes, cross-country skis, ice axes.

HUNTINGTON BEACH

PAT'S SKI AND SPORTS SHOPS   213-592-2821
8909 Adams Avenue   714-962-5521
Huntington Beach, California 92646
   Manufacturers' Seconds: 20 to 40 percent off list price on very
good brands. *Rent:* Sleeping bags, foam pads, tents, stoves, frame
packs, snowshoes, downhill skis. *Rental Sale:* Mid-September.
R.P.: Must be within 9-month period from time you rented the gear.

INDUSTRY

THE GRANT BOYS   213-965-8335
113 Puente Hills Mall
Industry, California 92670
   Semiannual Sales: Washington's Birthday and Halloween. *Discount:* 15 percent off suggested retail price, but on a limited range
of gear. *Discount to Nonprofit Organizations:* 10 percent discount.
Rent: Sleeping bags, tents, stoves, frame packs, lanterns, Coleman
lanterns, cots. *Rental Sales:* End of summer season.

KENTFIELD

TETONS WEST MOUNTAIN SHOP 415-457-8780
943 Sir Francis Drake Boulevard
Kentfield, California 94904

Semiannual Sales: September and April. *Discount:* 10 percent below manufacturers' suggested retail price on brands like Fabiano, Kelty, Camp 7, Snow Lion, Class 5. *Discount to Nonprofit Organizations:* 10 to 15 percent. *Manufacturers' Overruns and Seconds:* 20 percent discount. *Clearance Center:* 20 percent off list price. *Repairs:* Garments, bags, boots. *Rent:* Sleeping bags, tents, frame packs, soft packs, snowshoes, ice axes, crampons, cross-country skis. *Rental Sales:* September and April.

LA CANADA

SPORT CHALET 213-790-2717
920 Foothill Boulevard—Box 626
La Canada, California 91011

Discount: They do sell merchandise at a discount, but didn't indicate size of discount. They carry Nordica, Alpenlite, Class 5, Snow Lion, Woolrich, and other quality merchandise. *Discount to Nonprofit Organizations:* 10 to 25 percent discount to schools. *Manufacturers' Overruns and Seconds:* They do carry seconds and overruns but didn't specify the discount. *Repairs:* Skis, shoes, stoves. *Rent:* Sleeping bags, parkas, tents, hiking and climbing boots, stoves, frame packs, soft packs, snowshoes, downhill and cross-country skis, ice axes. *R.P.*

LA JOLLA

SAN DIEGO SKI CHALET 714-223-7173
7522 La Jolla Boulevard
La Jolla, California 92057

Manufacturers' Seconds: 20 percent discount. *Repairs:* Downhill skis. *Rent:* Sleeping bags, foam pads, tents, frame packs, soft packs, downhill skis, cross-country skis. *Rental Sales:* No set time—call. *R.P.*

LA PUENTE

DOUGHBOYS CLOTHING AND
CAMPING STORES   213-964-1259
17331 East Valley Boulevard
La Puente, California 91745
   Manufacturers' Overruns: 6 to 10 percent off list price.

LEMON GROVE

STANLEY ANDREWS SPORTS OUTFITTERS   714-460-6212
6941 Federal Boulevard
Lemon Grove, California 92045
   Discount to Nonprofit Organizations: 10 to 20 percent.
Rent: Sleeping bags, foam pads, tents, stoves, frame packs,
snowshoes, downhill and cross-country skis. *Rental Sales:* No set
time—call them. *R.P.*

LONG BEACH

SPORTS LIMITED   213-435-6521
1628 Long Beach Boulevard
Long Beach, California 90813
   Semiannual Sales: April and October. *Discount to Nonprofit
Organizations:* 10 percent. *Manufacturers' Overruns:* 50 percent
off list price on a limited variety of manufacturers' overruns. *Repairs:* Backpacks and all skis. *Rent:* Sleeping bags, tents, frame
packs, snowshoes, downhill and cross-country skis, ice axes, crampons. *Rental Sales:* End of each season. *R.P.*

LOS ANGELES

THE FAMOUS DEPARTMENT STORE   213-626-5151
530 South Main
Los Angeles, California 90013
   Discount: 5 to 10 percent below manufacturers' suggested list
price on brands like Coleman, Hirsch Weiss (White Stag), Sierra
Designs, Slumberjack, and Woolrich. *Manufacturers' Overruns and
Seconds:* 30 to 40 percent off regular price. *Rent:* Sleeping bags,
tents, frame packs, stoves, ice axes. *R.P.*

SPORTS LIMITED   213-627-6177

733 South Flower
Los Angeles, California 90017

Semiannual Sales: April and October. *Discount to Nonprofit Organizations:* 10 percent. *Repair:* Backpacks and skis. *Rent:* Sleeping bags, tents, frame packs, snowshoes, downhill and cross-country skis, ice axes, crampons. *Rental Sales:* No set time—check with them. *R.P.*

VALUE CENTERS   213-662-8131

3828 Sunset Boulevard
Los Angeles, California 90026

Discount: Limited variety of name brand merchandise (i.e. Coleman, Camp Ways, Wenzel) at 40 percent off list price.

WEST RIDGE SPORTS, INC.   213-820-5686

11930 W. Olympic Boulevard
Los Angeles, California 90064

Clearance Center: 10 to 40 percent discount on some excellent name brand merchandise. *Repairs:* Packs, clothing, skis. *Rent:* Sleeping bags, foam pads, tents, stoves, frame packs, soft packs, snowshoes, cross-country and downhill skis, ice axes, crampons.

WESTWOOD, INC.   213-474-6935

2043 Westwood Boulevard
Los Angeles, California 90025

Rent: Cross-country and downhill skis, foam pads, sleeping bags, parkas, tents, stoves, frame packs, soft packs, inflatable boats. *Rental Sales:* End of each season. *R.P.*

MAMMOTH LAKES

KITTREDGE SPORT SHOP   714-934-2423

P.O. Box 598—State Highway 203
Mammoth Lakes, California 93546

Discount: Some store brand merchandise (packs, sleeping bags, tents) are excellent buys, for the most part 20 to 25 percent less than comparable brand name merchandise. *Discount to Nonprofit Organizations:* 10 percent. *Clearance Center:* Up to 75 percent off list price. *Repairs:* Skis and archery equipment. *Rent:* Sleeping bags, foam pads, stoves, frame packs, soft packs, snowshoes, crampons, ice axes, downhill and cross-country skis. *Rental Sale:* October. *R.P.*

MAMMOTH INC. 714-934-6624
Highway 203
Mammoth Lakes, California 93546

Rent: Sleeping bags, foam pads, parkas, tents, stoves, frame packs, soft packs, downhill skis, inflatable boats. *Rental Sales:* End of each season. *R.P.*

MARINA DEL REY

BIKECOLOGY BIKE AND BACKPACK SHOPS 213-821-0766
4214 Lincoln Boulevard
Marina Del Rey, California 90291

Discount: 15 percent below manufacturers' suggested retail price on a few good brands, including Snow Lion. *Special Features:* They have an interesting "Anti-Rip-off Policy": If a bicycle is stolen during the first year, they'll sell you a replacement for their cost plus 10 percent.

SANDY'S SKI RENTALS 213-823-8133
4116 Lincoln Boulevard
Marina Del Rey, California 90291

Rent: Cross-country skis, sleeping bags, foam pads, parkas, tents, stoves, frame packs, soft packs, downhill skis, inflatable boats. *Rental Sales:* End of each season. *R.P.*

MERCED

DOWNHILL SKI & SPORT 209-383-1040
422 West 17 Street
Merced, California 95340

Semiannual Sales: March and April. *Clearance Center:* 50 percent savings. *Repairs:* Ski equipment. *Rent:* Frame packs, downhill skis. *Rental Sales:* They sell off their Alpine rental skis during the summer months.

MILL VALLEY

TETONS WEST MOUNTAIN SHOP 415-383-4050
87 East Blithdale Avenue
Mill Valley, California 94941

Semiannual Sales: September and April. *Discount:* 10 percent

An Improvised Shelter along the West Coast Trail
(Photo from the collection of North Country Mountaineering, Inc.)

below manufacturers' suggested list price on brands like Fabiano, Hine/Snowbridge, Kelty, Camp 7, Woolrich, Snow Lion, and Alpenlite. *Discount to Nonprofit Organizations:* 10 to 15 percent. *Manufacturers' Overruns and Seconds:* Some of the above brands at 20 percent below manufacturers' suggested list price. *Clearance Center:* 20 percent off list price. *Repairs:* They'll repair all garments and sleeping bags. *Rent:* Sleeping bags, tents, frame packs, soft packs, snowshoes, ice axes, crampons, cross-country skis. *Rental Sales:* September and April.

MODESTO

ROBBINS MOUNTAIN SHOP   209-529-6917
1508 10th Street
Modesto, California 95350
  Rent: Sleeping bags, foam pads, tents, stoves, frame packs, ice axes, crampons, canoes, cross-country skis. *Rental Sales:* End of winter and summer seasons.

MOUNT BALDY

MOUNT BALDY, INC.   714-9829023
Mount Baldy, California 91759
  Rent: Sleeping bags, foam pads, parkas, tents, stoves, frame packs, soft packs, downhill skis, inflatable boats. *Rental Sales:* End of each season. *R.P.*

NEWPORT BEACH

NEAL'S SPORTING GOODS   714-644-2121
27 Fashion Island
Newport Beach, California 92660
  Annual Sale: September. *Rent:* Sleeping bags, tents, frame packs, stoves, downhill skis. *R.P.:* One-half the rental fee may be applied toward the purchase price.

NORTHRIDGE

KELTY MOUNTAINEERING & BACKPACKING   213-993-0887
9066 Tampa
Northridge, California 91324

Semiannual Sales: April and October. *Repairs:* They repair only Kelty products. *Rent:* Sleeping bags, foam pads, tents, stoves, frame packs, soft packs, ice axes, cross-country skis. *Rental Sales:* April and October. *R.P.*

LITTLE STONE'S WILDERNESS
SHOPPE   214-886-9000
9545 Reseda Boulevard #13
Northridge, California 91324

Semiannual Sales: September and February. *Discount to Non-profit Organizations:* 10 percent. *Manufacturers' Overruns and Seconds:* 20 percent off list price. *Repairs:* They repair everything from holes in sleeping bags to broken zippers. A complete range of repairs. *Rent:* Sleeping bags, foam pads, tents, stoves, soft packs, frame packs, snowshoes, cross-country skis, ice axes, walkie-talkies. *Rental Sales:* September and February. *R.P.*

PACIFIC GROVE

BUGABOO MOUNTAINEERING   408-373-6433
170 Central Avenue
Pacific Grove, California 93950

Annual Sales: The first weekend in February. *Discount:* About 10 percent on selected items. *Clearance Center:* From 10 to 40 percent off on such brands as Pivetta, Galibier, Jansport, Trailwise, and Bugaboo products. *Repairs:* They repair all Bugaboo products.

PALO ALTO

THE NORTH FACE   415-327-1563
650 Quarry Road
Palo Alto, California 94304

Semiannual Sales: April and late September or early October. *Manufacturers' Seconds:* Their semiannual sale of seconds is held each April and late September or early October. You can expect about a 20 percent discount. *Clearance Center:* About 20 percent off list price—brands may include Kelty, North Face, Galibier, Sierra West, Nike shoes. *Repairs:* Cross-country skis. *Rent:* Sleeping bags, foam pads, tents, stoves, frame packs, soft packs, snowshoes, cross-country skis. *R.P.*

SIERRA DESIGNS   415-325-5231
217 Alma Street
Palo Alto, California 94301
   Repairs: Packs, sleeping bags, any Sierra Designs products.
Rent: Sleeping bags, foam pads, tents, frame packs, soft packs,
snowshoes, ice axes, cross-country skis, parkas, pots and pans.
Rental Sales: End of each season. *R.P.*

THE SKI HUT   415-321-2277
222 University Avenue
Palo Alto, California 94301
   Annual Sale: Early September. *Rent:* Down and synthetic fill
sleeping bags, tents, stoves, frame packs, soft packs, snowshoes,
downhill and cross-country skis, ice axes, crampons. *Rental
Sale:* Early September.

PINEDALE

ROBBINS MOUNTAIN SHOP   209-439-0745
7257 N. Abbey Road
Pinedale, California 93650
   Rent: Sleeping bags, foam pads, tents, stoves, frame packs, soft
packs, ice axes, crampons, canoes, cross-country skis. *Rental
Sales:* End of season.

REDDING

ALPINE OUTFITTERS   916-243-7333
1538 Market Street
Redding, California 96001
   Annual Sale: February. *Clearance Center:* 50 percent savings on
brands like Sierra Designs, Vasque, Fabiano, Kelty. *Repairs:* Skis.
They'll also stretch boots. *Rent:* Sleeping bags, foam pads, tents,
frame packs, snowshoes, downhill and cross-country skis, ice axes,
crampons. *R.P.*

REDLANDS

LITTLE STONE'S WILDERNESS SHOPPE   714-792-7777
218 Orange Street
Redlands, California 92373
  Semiannual Sales: March and April. *Discount to Nonprofit Organizations:* Contact them for their discount schedule. *Repairs:* They repair skis and soft goods.

REDONDO BEACH

PAT'S SKI AND SPORTS SHOPS   213-378-8551
115 Palos Verdes Boulevard
Redondo Beach, California 90277
  Semiannual Sales: They have a pre-season sale in March or April. Their summer clearance sale is held each August. *Manufacturers' Overruns and Seconds:* 20 to 40 percent discount on name brand merchandise. *Rent:* Sleeping bags, tents, foam pads, stoves, frame packs, snowshoes, downhill skis. *Rental Sale:* Mid-September. *R.P.:* Within nine months of the rental.

ROSEMEAD

A-C RENTALS   213-288-4435
8702 Valley
Rosemead, California 91770
  Discount to Nonprofit Organizations: 10 percent discount. *Manufacturers' Overruns and Seconds:* 20 percent off list price. They didn't specify the brands, but generally they carry a limited range of equipment, including Coleman. *Repairs:* They repair Coleman products (hard line only). *Rent:* Sleeping bags, foam pads, tents, stoves, frame packs, soft packs, canoes.

DOUGHBOYS CLOTHING AND
CAMPING STORES   213-877-5411
8334 Garvey
Rosemead, California 91770
  Discount: 3 to 5 percent below manufacturers' suggested list price on brands that include Slumber Jack, Washington Quilt, Wenzel, Stan Sport, Sports Master. *Manufacturers' Overruns and Sec-*

onds: They carry these sometimes, at approximately a 6 to 8 percent discount.

SACRAMENTO

ALPINE WEST   916-441-1627
1021 R
Sacramento, California 95814
   Rent: Tents, sleeping bags, foam pads, stoves, frame packs, snowshoes, crampons, cross-country skis, gaiters. *R.P.*

ANTELOPE CAMPING EQUIPMENT   916-489-9591
1621 Fulton Avenue
Sacramento, California 95825
   Discount: 10 percent discount on cash sales over $200, 15 percent discount on cash sales over $300. *Manufacturers' Seconds:* 20 percent below list price—an excellent place to check for seconds of Antelope backpacks. *Repairs:* Packs. *Rent:* Sleeping bags, foam pads, tents, frame packs, snowshoes, cross-country skis. *Rental Sale:* Winter—check with them for this year's date. *R.P.*

GOODWIN-COLE SPORTS   916-452-6641
1315 Alhambra Boulevard
Sacramento, California 95816
   Semiannual Sales: Spring and fall. *Discount to Nonprofit Organizations:* Contact them for discount schedule for quantity purchases. *Repairs:* Boots and sportshoe repairs, including resoling. They'll also do all sewing repairs. They're a Coleman repair center, and they repair rafts. *Rent:* Sleeping bags, foam pads, tents, stoves, frame packs, snowshoes, downhill and cross-country skis. *R.P.*

MCINTOSH'S SPORTS CENTER   916-488-7181
4120 El Camino
Sacramento, California 95821
   Annual Sale: September 1. *Discount to Nonprofit Organizations:* 10 percent. *Repairs:* Skis, tennis rackets. *Rent:* Frame packs, downhill skis, kayaks, cross-country skis. *Rental Sales:* Every two years. *R.P.*

MCINTOSH'S SPORTS COTTAGE   916-456-3975
4768 J Street
Sacramento, California 95823

Annual Sale: September 1. *Discount to Nonprofit Organizations:* 10 percent. *Repairs:* Skis, tennis rackets. *Rent:* Frame packs, downhill skis, kayaks, cross-country skis. *Rental Sales:* Every two years. *R.P.*

SIERRA OUTFITTERS   916-481-2480
2903 Fulton
Sacramento, California 95821
   Rent: Sleeping bags, foam pads, tents, frame packs, soft packs, snowshoes, ice axes, crampons, downhill and cross-country skis. *R.P.*

SAN DIEGO

A16 WILDERNESS CAMPING OUTFITTERS   714-283-2374
4620 Alvarado Canyon Road
San Diego, California 92120
   Annual Sale: Mother's Day. *Discount to Nonprofit Organizations:* They'll give a discount to schools only. Check with them for the amount. *Manufacturers' Overruns and Seconds:* 20 to 80 percent off list price. Obviously, this is a great place to look for seconds of their own brand of gear. *Repairs:* A16 products. *Rent:* Sleeping bags, foam pads, tents, frame packs, snowshoes, cross-country skis. *Rental Sale:* Every Mother's Day. *Special Features:* Their annual Mother's Day Sale coincides with their annual Swap Meet, to which anyone can bring used camping gear for trade or sale.

GREGORY MOUNTAIN PRODUCTS   714-284-4050
REPAIR AND DESIGN
4620 Alvarado Canyon Road
San Diego, California 92120
   Repairs: They will repair all backpacking equipment, except boots. They'll also properly clean both down and synthetic garments and sleeping bags.

SAN DIEGO SKI CHALET   714-459-2691
4004 Sports Arena Boulevard
San Diego, California 93401
   Manufacturers' Seconds: They have these occasionally, at a 20 percent discount. *Repairs:* Downhill skis. *Rent:* Sleeping bags, foam pads, tents, frame packs, soft packs, downhill and cross-country skis. *Rental Sales:* No set dates—call them. *R.P.*

STANLEY ANDREWS SPORTS
OUTFITTERS   714-236-9191
840 B Street
San Diego, California 92101
   Discount to Nonprofit Organizations: 10 to 20 percent.
Rent: Sleeping bags, foam pads, tents, stoves, frame packs,
snowshoes, downhill and cross-country skis. *Rental Sales:* No set
date—ask them. *R.P.*

STANLEY ANDREWS SPORTS
OUTFITTERS   714-276-1311
3988 Clairemont Mesa Boulevard
San Diego, California 92117
   Discount to Nonprofit Organizations: 10 to 20 percent.
Rent: Sleeping bags, foam pads, tents, stoves, frame packs,
snowshoes, downhill and cross-country skis. *Rental Sales:* No set
dates—ask them. *R.P.*

STANLEY ANDREWS SPORTS
OUTFITTERS   714-232-7362
443 12th Street
San Diego, California 92101
   Rent: Sleeping bags, foam pads, tents, stoves, frame packs,
snowshoes, downhill and cross-country skis. *Rental Sales:* No set
dates—ask them. *R.P.*

SAN FRANCISCO

CALIFORNIA SURPLUS SALES   415-861-1083
1107 Mission
San Francisco, California 94103
   Discount: They sell general camping goods at 15 to 20 percent
discount. *Discount to Nonprofit Organizations:* 15 percent. *Manufacturers' Overruns and Seconds:* 30 to 40 percent discount.

G & M SALES CO.   415-863-2855
1667 Market
San Francisco, California 94102
   Semiannual Sales: At the times seasons change—check with
them for more specific details. *Discount:* They wouldn't tell us how
much they discount, but they do. They carry brands that include
Raichle, Gerry, Trailwise, Camp Trails, Comfy, and Summit. *Manufacturers' Overruns and Seconds:* They wouldn't tell us which

brands, or what the discount is. If you're in the area, it might be worth your while to stop by. *Repairs:* Cross-country skis, downhill skis, packs, stoves, lanterns, and some tent repairs. *Rent:* Sleeping bags, foam pads, parkas, tents, stoves, frame packs, soft packs, snowshoes, downhill and cross-country skis, lanterns, ice chests, ski clothing. *Rental Sales:* End of season. *R.P.*

MOUNTAIN SHOP, INC.   415-362-8477
228 Grant Avenue
San Francisco, California 94108
 Rent: Sleeping bags, tents, frame packs, soft packs, snowshoes, cross-country skis. *Rental Sale:* Mid-October. *R.P.*

THE NORTH FACE   415-665-6044
Stonestown Shopping Center
San Francisco, California 94132
 Semiannual Sales: April and September. These are sales of North Face seconds at 20 to 30 percent off list price. *Repairs:* Only North Face equipment. *Rent:* Sleeping bags, foam pads, tents, stoves, frame packs, soft packs. *R.P.*

THE SMILIE COMPANY   415-421-2459
575 Howard Street
San Francisco, California 94104
 Rent: Sleeping bags, foam pads, tents, stoves, frame packs, soft packs, snowshoes, ice axes, crampons, cross-country skis, pot kits. *Rental Sales:* October and March. *R.P.*

SAN GABRIEL

ALTA SPORT   213-287-0736
9034 Huntington Drive
San Gabriel, California 91775
 Annual Sale: March. *Repairs:* Skis. *Rent:* Sleeping bags, foam pads, frame packs, snowshoes, downhill skis.

SAN JOSÉ

FREEMANS SPORT CENTERS   408-244-7300
840 Town & Country Village
San José, California 95218
 Semiannual Sales: Each September and in the spring. *Discount to Nonprofit Organizations:* 10 percent. *Repairs:* Alpine and Nordic

ski bindings, also tennis and archery repairs. *Rent:* Sleeping bags, frame packs, snowshoes, downhill and cross-country skis, ice axes, crampons. *Rental Sale:* Usually each fall. *R.P. Special Features:* They offer a discount on rentals to nonprofit organizations.

SAN LUIS OBISPO

GRANITE STAIRWAY MOUNTAINEERING 805-541-1533
871 Santa Rosa
San Luis Obispo, California 93401
  Semiannual Sales: Usually April and October (up to 40 percent off). *Discount to Nonprofit Organizations:* 10 percent. *Manufacturers' Seconds:* 25 to 40 percent off list price. *Rent:* Sleeping bags, parkas, tents, stoves, frame packs, soft packs, snowshoes, downhill and cross-country skis, ice axes, crampons, ski racks. *Rental Sales:* Only damaged equipment. *R.P.*

SANTA ANA

HOLUBAR 714-549-8541
3650 South Bristol Street
Santa Ana, California 92704
  Manufacturers' Seconds: 10 to 20 percent off list price. Look for seconds of their own brand of merchandise. *Rent:* Sleeping bags, foam pads, tents, stoves, frame packs, soft packs, snowshoes, cross-country skis, ice axes. *Rental Sales:* End of season. *R.P.*

NEAL'S SPORTING GOODS 714-547-5723
219 East 4th
Santa Ana, California 92701
  Annual Sale: September. *Rent:* Sleeping bags, tents, frame packs, stoves, downhill skis. *R.P.:* One-half the rental fee can be applied toward the purchase price.

PAT'S SKI & SPORTS SHOP 714-834-1006
2239 North Tustin Avenue
Santa Ana, California 92701
  Semiannual Sales: Pre-season sales are usually in March or April. Their summer clearance sale is held each August. *Manufacturers' Overruns and Seconds:* 20 to 40 percent off list price. They didn't specify the brands they carry in overruns and seconds, but their regular merchandise includes High and Light,

North Face, Gerry, Jansport. *Rent:* Sleeping bags, foam pads, tents, stoves, frame packs, snowshoes, downhill skis. *Rental Sale:* Mid-September. *R.P.:* Merchandise must be bought within nine months of the rental in order for the rental fee to be applied toward purchase price.

SANTA BARBARA

GRANITE STAIRWAY MOUNTAINEERING   805-682-1083
2040 State Street
Santa Barbara, California 93105
  Semiannual Sales: April and October, up to 40 percent off list price during the sales only. *Discount to Nonprofit Organizations:* 10 percent, upon approval only. Speak with them about it. *Manufacturers' Overruns and Seconds:* 20 to 40 percent off list price. *Repairs:* Skis, packs, bags, all repairs. *Rent:* Sleeping bags, parkas, tents, stoves, frame packs, soft packs, snowshoes, downhill skis, ice axes, crampons, cross-country skis, ski racks. *Rental Sales:* Only damaged equipment. *R.P.*

SANTA FE SPRINGS

OXMAN'S SURPLUS   213-921-1106
14128 East Rosecrans Avenue
Santa Fe Springs, California 90670
  Discount: 5 to 10 percent below list price on a limited range of brands such as Slumberjack and Stansport. *Discount to Nonprofit Organizations:* 5 percent. *Manufacturers' Overruns and Seconds:* 20 percent off list price.

SANTA MONICA

BIKECOLOGY, INC.   213-829-7681
2910 Nebraska Avenue
Santa Monica, California 90406
  Discount: 15 percent below manufacturers' suggested list price on some items. *Repairs:* All bicycle repairs. *Special Features:* They have an interesting "Anti-Rip-off Policy": If a bicycle is stolen during the first year, they'll sell you a replacement for their cost plus 10 percent.

BIKECOLOGY, INC.   213-828-6053
3006 Wilshire Boulevard
Santa Monica, California 90405
   Discounts: 15 percent below manufacturers' suggested list price
on some items. *Repairs:* All bicycle repairs. *Special Features:* See
details of the firm's "Anti-Rip-off Policy" above.

SANTA ROSA

THE BACKPACKERS TENT   707-544-2040
533 5th Street
Santa Rosa, California 95401
   Discount to Nonprofit Organizations: Varies with the amount of
purchases. Boy Scouts get a flat 10 percent discount. *Manufacturers' Seconds:* 15 to 25 percent discount on seconds of name brand
merchandise. *Rent:* Frame packs, soft packs, snowshoes. *Rental
Sales:* Twice a year—check with them. *R.P.*

SAUGUS

CANYON COUNTRY, INC.   805-251-5030
20617 Soledad Canyon Road
Saugus, California 91351
   Rent: Sleeping bags, foam pads, parkas, tents, stoves, frame
packs, soft packs, downhill skis, inflatable boats. *Rental Sales:* End
of each season. *R.P.*

SONORA

SONORA MOUNTAINEERING   209-532-5621
171 N. Washington Street
Sonora, California 95370
   Rent: Sleeping bags, tents, frame packs, snowshoes, cross-country skis. *R.P.*

SOUTH GATE

DICK CEPEK   213-569-1675
9201 California Avenue
South Gate, California 90280
   Semiannual Sales: June and January. *Discount:* 30 percent off

list price on an odd assortment of gear. *Manufacturers' Overruns and Seconds:* 50 percent off list price on serviceable but not fancy (i.e., name brands) equipment. *Clearance Center:* 50 percent off list price.

SOUTH LAKE TAHOE

THE OUTDOORSMAN   916-541-1660
Box 8877
South Lake Tahoe, California 95731
   Semiannual Sales: Labor Day and Washington's Birthday. *Repairs:* Minor repairs. *Rent:* Snowshoes, downhill and cross-country skis. *Rental Sales:* Ski equipment only, in the spring. Call them to check the date.

TATUM OUTFITTERS   916-544-1933
3131 Harrison Avenue
South Lake Tahoe, California 95705
   Annual Sale: After Christmas. *Discount:* Some equipment sold at less than manufacturers' suggested retail price. Discounts start at 5 percent. *Repairs:* Minor repairs on all equipment. *Rent:* Sailboats, downhill skis. *Rental Sales:* End of the season. *R.P.*

STOCKTON

KELMOORE MOUNTAIN SPORTS   209-466-6620
409 West Fremont Street
Stockton, California 95203
   Semiannual Sales: End of April and end of July. *Discount to Nonprofit Organizations:* 10 to 15 percent. *Repairs:* Backpacks, stoves, cross-country ski equipment. *Rent:* Sleeping bags, foam pads, tents, stoves, frame packs, soft packs, kayaks, ice axes, crampons, cross-country skis, river rafts. *Rental Sales:* Contact them. *R.P.*

SUNNYVALE

FREEMANS SPORT CENTERS   408-732-3300
711 Town & Country Village
Sunnyvale, California 94086
   Semiannual Sales: Usually each September and in the spring.

Manufacturers' Overruns and Seconds: They occasionally bring these in for special sales, at 20 to 30 percent discount. *Repairs:* Alpine and Nordic ski bindings. *Rent:* Sleeping bags, foam pads, snowshoes, downhill and cross-country skis, ice axes, crampons. *Rental Sales:* Usually each fall. *R.P. Special Features:* They offer a discount on rentals to nonprofit organizations.

SYLMAR

AA RENTAL CENTER   213-365-7168
13245 N. Maclay
Sylmar, California 91342
   Rent: Sleeping bags, tents, stoves, pack bags, lanterns, coolers, other miscellaneous supplies.

TARZANA

GRANITE STAIRWAY MOUNTAINEERING   213-345-4266
5425 Reseda Boulevard
Tarzana, California 91356
   Discount for Nonprofit Organizations: 10 percent. *Manufacturers' Overruns and Seconds:* 20 to 40 percent off list price. *Repairs:* All repairs. *Rent:* Sleeping bags, parkas, tents, stoves, frame packs, soft packs, snowshoes, downhill and cross-country skis, ice axes, crampons, ski racks. *Rental Sales:* Only damaged equipment. *R.P.*

SANDY'S SKI & SPORTS RENTALS   213-345-5775
19539 Ventura Boulevard
Tarzana, California 91356
   Rent: Sleeping bags, foam pads, parkas, tents, stoves, frame packs, soft packs, downhill skis, inflatable boats. *Rental Sales:* End of season. *R.P.*

TORRENCE

THE SKI RACQUET   213-371-3533
20611 Hawthorne Boulevard
Torrence, California 90503

Manufacturers' Overruns: 10 to 20 percent off list price. *Repairs:* Skis and tennis rackets. *Rent:* Sleeping bags, foam pads, tents, stoves, frame packs, snowshoes, ice axes, downhill and cross-country skis, ski racks, water skis. *R.P.*

TORRENCE INC.   213-371-5932
21004 Hawthorne Boulevard
Torrence, California 90503
   Rent: Sleeping bags, foam pads, parkas, tents, stoves, frame packs, soft packs, downhill skis, inflatable boats. *Rental Sales:* End of each season. *R.P.*

UPLAND

PACK & PITON   714-982-7408
1252 West Foothill
Upland, California 91786
   Manufacturers' Seconds: 20 to 30 percent off list price. They don't specify which brand(s) of seconds they carry, but their merchandise is generally first-rate (Sierra Designs, North Face, Kelty, Hine/Snowbridge). *Rent:* Sleeping bags, tents, frame packs, soft packs, snowshoes, ice axes, crampons. *Rental Sale:* Second weekend after Labor Day. *R.P.*

VAN NUYS

SANDY'S SKI RENTALS   213-893-4211
15331 Roscoe Boulevard
Van Nuys, California 91401
   Rent: Sleeping bags, foam pads, parkas, tents, stoves, frame packs, soft packs, downhill skis, inflatable boats.

VALUE CENTERS   213-787-0778
6255 Sepulveda Boulevard
Van Nuys, California 91401
   Discount: They always sell their merchandise at less than manufacturers' suggested list price. Discounts range from 5 to 40 percent on a decent but limited range of brands: Stan Sport, Coleman, Camp Way, Himalayan Industries, Wenzel. *Manufacturers' Overruns:* A limited variety, but at huge savings: 50 percent. *Rental Sales:* End of each season. *R.P.*

WEST COVINA

ALPINE COUNTRY   213-962-4311
1030 West Covina Parkway
West Covina, California 91790

  Semiannual Sales: May and October. *Discount to Nonprofit Organizations:* 20 percent—a very good nonprofit discount. *Manufacturers' Overruns and Seconds:* 30 percent off list price. Do check it out; they carry some excellent brands. *Repairs:* Tents, packs, clothing, sleeping bags. *Rent:* Sleeping bags, foam pads, parkas, tents, stoves, frame packs, soft packs, snowshoes, ice axes, crampons, cross-country skis. *Rental Sale:* October. *R.P.*

WEST LOS ANGELES

A16 WILDERNESS CAMPING OUTFITTERS   213-473-4574
11161 West Pico Boulevard
West Los Angeles, California 90025

  Annual Sale: The Sunday nearest to Halloween. *Repairs:* They repair their own products. *Rent:* Sleeping bags, foam pads, tents, frame packs, snowshoes, cross-country skis. *Rental Sale:* They sell their rental equipment on Mother's Day each year. *R.P. Special Features:* They invite customers to come to their annual Swap Meet, held on Mother's Day each year. Bring your used equipment for sale or trade.

WEST RIDGE SPORTS   213-820-5686
11930 West Olympic Boulevard
West Los Angeles, California 90064

  Clearance Center: 10 to 40 percent off on brands that may include Jansport, Fabiano, Raichle, Pivetta, Kelty, Twin Peaks, Sierra Designs. *Repairs:* Packs, clothing, skis. *Rent:* Sleeping bags, foam pads, tents, stoves, frame packs, snowshoes, downhill and cross-country skis, ice axes, crampons. *Rental Sale:* Periodically— ask them if there's a rental sale coming up.

WESTMINSTER

LONG BEACH SURPLUS   714-892-8306
7722 Garden Grove
Westminster, California 92683
  Annual Sale: End of summer—ask them for this year's dates.
Discount: They have unadvertised specials during which brands like
Alpine Designs, Mountain Products, Ascente, and Class 5 are sold
at about 30 percent less than list price. *Discount to Nonprofit
Organizations:* 10 to 15 percent. *Manufacturers' Overruns and
Seconds:* 5 to 10 percent less than list price. *Clearance Center:*
Discounts of 10 to 20 percent. *Repairs:* Coleman and Optimus
stoves.

SANDY'S SKI RENTALS   714-893-3585
15352 Beach Boulevard
Westminster, California 92683
  Rent: Sleeping bags, foam pads, parkas, tents, stoves, frame
packs, soft packs, downhill skis, inflatable boats. *Rental Sales:* End
of each season. *R.P.*

WOODLAND HILLS

SPORTS LTD.   213-346-3330
22642 Ventura
Woodland Hills, California 91362
  Semiannual Sales: April and October. *Discount to Nonprofit
Organizations:* 10 percent. *Repairs:* Backpacks and all skis.
Rent: Sleeping bags, tents, frame packs, snowshoes, downhill and
cross-country skis, ice axes, crampons. *Rental Sales:* Seasonally.
R.P.

YOSEMITE

THE MOUNTAIN SHOP   209-372-4611 Ext. 296
Curry Village
Yosemite National Park, California 95389
  Manufacturers' Overruns: 25 percent off list price. These ought
to be interesting, as the Mountain Shop carries a small line of first-
rate products. *Rent:* Foam pads, boots, frame packs, soft packs,
snowshoes, downhill and cross-country skis, bicycles, ice axes,
crampons. All rentals are through the Yosemite Mountaineering

School. *Special Features:* They are affiliated with two other shops in Yosemite National Park—Village Sports Shop, and Badger Pass Ski Shop.

The Boulder Mountaineer Rental Department
(Photo courtesy of Bob Culp)

ASPEN

BOULDER MOUNTAINEER IN ASPEN   303-925-2849
Suite 204
315 Hyman Avenue
Aspen, Colorado 81611
   Rent: To the best of our knowledge, this is the largest rental service of its type in the United States. Bob Culp is now devoting 1000 feet to rental space and is offering: sleeping bags, foam pads, parkas, tents, hiking boots, technical climbing boots, stoves, frame packs, soft packs, ice axes, crampons, cross-country skis, ski mountaineering skis, Jumar ascenders, vertical ice tools (ice screws, rigid crampons, North Wall hammers, Terrordactyls). They also carry, for rent, all technical rock climbing hardware: hard hats, rock climbing hammers, pitons, carabiners, nuts, bivouac hammocks, and other Big Wall gear. *R.P.*

AURORA

AURORA SALVAGE & SURPLUS   303-366-3002
1556 Florence Street
Denver, Colorado 80010
   Discount: 10 to 30 percent less than manufacturers' suggested retail price on brands like Dunham and Camp Ways. *Discount to Nonprofit Organizations:* 10 percent. *Manufacturers' Overruns and Seconds:* 20 to 50 percent off list price. *Clearance Center:* 20 to 50 percent off list price. *Repairs:* Tent poles, electrical equipment. *Special Features:* They carry caving supplies. Aurora, by the way, is two miles east of Denver.

BOULDER

BOULDER MOUNTAINEER   303-443-8355
1329 Broadway
Boulder, Colorado 80302
   Discount: 30 percent discount on discontinued equipment. *Repairs:* Cross-country ski repairs. *Special Features:* They will cus-

tom modify your personal gear. For example, they'll alter pant and parka length, and sleeve length.

HOLUBAR
303-449-1731
1975 North 30th Street
Boulder, Colorado 80302

Semiannual Sales: Winter and summer—call them for more specific dates. *Discounts for Nonprofit Organizations:* 10 to 30 percent. The latter is one of the heftiest discount structures for nonprofit groups we've seen. *Manufacturers' Overruns and Seconds:* 10 to 40 percent off list price. Obviously, this is a great place to check for seconds and overruns of Holubar's own brands at substantial savings! *Repairs:* They repair everything they sell. *Rent:* Sleeping bags, foam pads, tents, frame packs, soft packs, snowshoes, ice axes, cross-country skis.

MADDEN MOUNTAINEERING LTD.
303-442-1982
P.O. Box 3206
1840 Commerce
Boulder, Colorado 80301

Discount: 10 percent on orders over $100, 15 percent on orders over $250, 20 percent discount for orders over $500. *Discount to Nonprofit Organizations:* Same discount schedule as the regular discount schedule.

MOUNTAIN SPORTS
303-443-6770
821 Pearl
Boulder, Colorado 80302

Manufacturers' Seconds: 10 to 20 percent savings on a limited range of very good brands. *Repairs:* They will repair any lightweight camping equipment. *Rent:* Sleeping bags, foam pads, tents, frame packs, snowshoes, cross-country skis. *R.P.*

BRECKENRIDGE

RECREATION SPORTS
303-453-2194
Box 1190
Breckenridge, Colorado 80424

Annual Sales: Usually during April. *Clearance Center:* 20 to 30 percent off list price on brands that may include Kelty, Camp 7, Nordica, Raichle, Dunham, Sport Obermeyer, Vasque Voyageur, Pivetta, and Jansport. *Repairs:* Skis, tennis rackets, bicycles.

Rent: Sleeping bags, tents, frame packs, downhill and cross-country skis, bicycles. *Rental Sale:* May. *R.P.*

COLORADO SPRINGS

HOLUBAR MOUNTAINEERING, LTD.   303-634-5261
1776 West Uintah
Colorado Springs, Colorado 80904
   Semiannual Sales: Winter and summer—call them to find out the exact dates. *Discount for Nonprofit Organizations:* The manager of this Holubar store didn't indicate their nonprofit discount structure on our questionnaire. If you encounter problems, you may want to contact the Boulder store, where they'll sell you the same merchandise at from 10 to 30 percent off list price. *Manufacturers' Overruns and Seconds:* Check for seconds of their own brand of merchandise. Savings should range from 10 to 40 percent, as at the Boulder store. *Repairs:* They repair everything they sell. *Rent:* Sleeping bags, foam pads, tents, frame packs, soft packs, snowshoes, ice axes, cross-country skis.

MOUNTAIN CHALET   303-633-0732
226 N. Tejon
Colorado Springs, Colorado 80902
   Rent: Sleeping bags, foam pads, tents, frame packs, soft packs, kayaks, cross-country skis. *Rental Sale:* February. *R.P.*

DENVER

BACKWOODS AND MOUNTAIN CHALET   303-377-2783
938 South Monaco
Denver, Colorado 80224
   Semiannual Sales: Early spring and early fall. *Discount to Nonprofit Organizations:* Only on sales of kits. *Manufacturers' Seconds:* They carry them sometimes. *Repairs:* Stoves and some packs. *Rent:* Sleeping bags, foam pads, tents, stoves, frame packs, soft packs, snowshoes, ice axes, crampons, cross-country skis.

DAVE COOK SPORTING GOODS   303-892-1929
16th and Market
Denver, Colorado 80202
   Annual Sale: Each fall. *Discount:* They do sell at less than manu-

facturers' list price, but didn't specify what brands, or at how much of a discount. *Discount to Nonprofit Organizations:* They do give them but didn't state the percentage. *Manufacturers' Overruns and Seconds:* 20 to 30 percent off list price. This ought to be very interesting, as Dave Cook's carries an excellent line of top brands. *Clearance Center:* 50 to 60 percent off list price on a range of merchandise that may at times include such prestigious brands as Class 5, Gerry, Alpenlite, Alpine Designs, and Jansport. *Repairs:* Ski equipment. *Rent:* Snowshoes, downhill and cross-country skis. *Rental Sale:* Each fall.

EASTERN MOUNTAIN SPORTS   303-571-1160
1428 15th Street
Denver, Colorado 80202

Semiannual Sales: October and April. *Discount:* They occasionally sell merchandise at less than list price. Their own brand of merchandise is generally 10 to 20 percent less than comparable brand name gear. *Discount to Nonprofit Organizations:* Up to 20 percent. *Manufacturers' Overruns and Seconds:* They do carry them, but the amount of discount varies. *Repairs:* Some repairs "when we can find someone to do it." *Rent:* Sleeping bags, foam pads, stoves, tents, frame packs, soft packs, snowshoes, cross-country skis. *Rental Sales:* October and April. *R.P.*

FORREST MOUNTAIN SHOPS   303-477-1722
1517 Platte Street
Denver, Colorado 80202

Semiannual Sales: Times of sales vary each year. Check with them. *Discount to Nonprofit Organizations:* They do give them, but didn't specify how much. *Manufacturers' Seconds:* They have their own seconds at a 40 percent discount—an extraordinary discount on an extraordinary line of equipment. *Rent:* Foam pads, soft packs, snowshoes, ice axes, crampons. *Rental Sale:* Late spring. *R.P.*

HOLUBAR MOUNTAINEERING, LTD.   303-758-6366
2490 South Colorado
Denver, Colorado 80222

Semiannual Sales: Winter and summer—call them for more specific dates. *Discount to Nonprofit Organizations:* 10 to 30 percent. The latter is one of the heftiest discount structures we've seen. *Manufacturers' Overruns and Seconds:* 10 to 40 percent off list price. Obviously, this is a great place to check for Holubar's own brands at substantial savings. *Repairs:* They repair everything they sell. *Rent:* Sleeping bags, foam pads, tents, frame packs, soft packs, snowshoes, ice axes, cross-country skis.

TELEMARK SKI & MOUNTAIN SPORTS   303-837-1260
416 East 7th Avenue
Denver, Colorado 80203
   Semiannual Sales: August and April. *Discount to Nonprofit Organizations:* They discount to some organizations. *Manufacturers' Overruns and Seconds:* They have them occasionally. Look for them if you stop by the store. *Repairs:* Skis and boots. They'll do sewing repairs upon request. *Rent:* Sleeping bags, foam pads, parkas, tents, stoves, frame packs, soft packs, snowshoes, cross-country and downhill skis, ice axes, crampons. *Rental Sales:* September (backpacking equipment) and May (ski equipment). *R.P.:* Sometimes.

ESTES PARK

COLORADO MOUNTAIN SPORTS   303-586-2829
237 West Elkhorn Avenue
Estes Park, Colorado 80517
   Sales: August, October, February. *Discount to Nonprofit Organizations:* 10 percent. *Manufacturers' Overruns and Seconds:* 20 to 50 percent off list price. *Clearance Center:* 40 percent discount. *Rent:* Sleeping bags, foam pads, tents, stoves, frame packs, soft packs, snowshoes, ice axes, crampons, cross-country skis, kayaks. *Rental Sale:* October. *R.P.*

FORT COLLINS

HOLUBAR STORES   303-484-2872
2715 South College Avenue
Fort Collins, Colorado 80521
   Semiannual Sales: Winter and summer—call them for more specific dates. *Discount to Nonprofit Organizations:* 10 to 30 per cent. *Manufacturers' Overruns and Seconds:* 10 to 40 per cent off list price—a good place to check for Holubar's own brands at substantial savings. *Repairs:* They repair everything they sell. *Rent:* Sleeping bags, foam pads, tents, frame packs, soft packs, snowshoes, ice axes, cross-country skis.

Long Peak Trail
(Photo from the collection of North Country Mountaineering, Inc.)

THE MOUNTAIN SHOP   303-493-5720
126 West Laurel
Fort Collins, Colorado 80521

Semiannual Sales: Spring and fall—call them for more specific information. *Clearance Center:* 10 to 40 percent off list price. Brands may range from Galibier, Kelty, North Face, and Wilderness Experience, to Snow Lion and Sierra Designs. *Repairs:* Cross-country skis and some soft goods. *Rent:* Sleeping bags, foam pads, tents, frame packs, soft packs, snowshoes, ice axes, crampons, cross-country skis. *Rental Sale:* End of season. *R.P.*

GRAND JUNCTION

MARMOT MOUNTAIN WORKS, LTD.   800-525-4224
P.O. Box 2433   300-243-2339
Grand Junction, Colorado 81501

Special Features: They have a thriving mail order business, with fairly hefty prices. But 5 percent of what you pay will be donated to the environmental conservation group of your choice.

GRAND LAKE

GOOD EARTH MOUNTAIN SHOP
907 Grand Avenue (Box 452)
Grand Lake, Colorado 80447

Discount: They sometimes offer a 10 to 20 percent discount. *Manufacturers' Seconds:* 5 percent off list price. *Repairs:* Stoves and garments. They're especially adept at snaps and tears. *Rent:* Sleeping bags, foam pads, parkas, tents, climbing boots, hiking boots, stoves, frame packs, soft packs, snowshoes, canoes, cartop carriers, bicycles, ice axes, crampons, cross-country skis. *R.P.*

RECREATION RENTALS   303-627-3642
Box K
Grand Lake, Colorado 80447

Discount: They sometimes sell at less than suggested list price. To quote Wendell Funk, the store's manager and buyer, "We consider any dealer should sell at any price he chooses. We don't believe in 'equalized robbery' (alias suggested list price)." *Manufacturers' Seconds:* 5 to 10 percent off list price. *Repairs:* Stoves, garments, snaps and tears. *Rent:* Sleeping bags, foam pads, parkas, tents, climbing boots, hiking boots, stoves, frame packs, soft packs, snowshoes, canoes, cartop carriers, bicycles, ice axes, crampons, cross-country skis. *R.P.*

GREELEY

COLORADO MOUNTAIN SPORTS   303-356-7873
1982 Greeley Mall
Greeley, Colorado 80631
   Sales: August, October, February. *Manufacturers' Overruns and Seconds:* 20 to 50 percent off list price. *Clearance Center:* 40 percent off list price for brands like Dunham, Dexter, Sierra Designs, and Camp 7. *Rent:* Sleeping bags, foam pads, tents, stoves, frame packs, soft packs, snowshoes, ice axes, crampons, cross-country skis, kayaks. *Rental Sale:* October. *R.P.*

GUNNISON

CARROL'S LTD.   303-641-1127
125 North Main Street
Gunnison, Colorado 81230
   Discount to Nonprofit Organizations: 10 percent, but not for all nonprofit groups. *Manufacturers' Seconds:* 10 to 25 percent off list price. *Clearance Center:* 10 to 25 percent off list price on brands like Alpenlite, Vasque Voyageur, Lowa, Hine/Snowbridge, and North Face. *Repairs:* Skis. *Rent:* Cross-country skis. *Rental Sales:* March and April. *R.P.*

LAKEWOOD

CHRISTY SPORTS   303-237-3483
9885 West Colfax Avenue
Lakewood, Colorado 80215
   Discount to Nonprofit Organizations: 20 percent. *Repairs:* Skis and backpacks. *Rent:* Tents, stoves, frame packs, snowshoes, downhill and cross-country skis, ice axes, crampons. *Rental Sale:* March.

MOUNTAIN WORLD LTD.   303-233-7651
2640 Youngfield
Lakewood, Colorado 80215
   Semiannual Sales: March and September. *Discounts to Non-profit Organizations:* 10 to 20 percent. *Repairs:* Cross-country skis and equipment. *Rent:* Sleeping bags, foam pads, tents, frame packs, soft packs, snowshoes, ice axes, crampons, cross-country skis. *Rental Sale:* Second week of March. *R.P.*

PUEBLO

MOUNTAIN CHALET   303-545-9890
140 West 29 Street
Pueblo, Colorado 81008
   Semiannual Sales: They hold them but didn't tell us when. Call them for further information. *Rent:* Sleeping bags, tents, stoves, frame packs, soft packs, foam pads, ice axes, cross-country skis, kayaks. *Rental Sale:* February. *R.P.*

STEAMBOAT SPRINGS

MOUNTAINCRAFT INCORPORATED   303-879-2368
810 Lincoln Avenue
P.O. Box 359
Steamboat Springs, Colorado 80477
   Discount to Nonprofit Organizations: Varies. *Repairs:* Bicycles and cross-country skis. *Rent:* Bicycles, cross-country skis. *R.P.*

VAIL

CHRISTY SPORTS   303-476-2244
Box 1388
307 East Bridge
Vail, Colorado 81657
   Semiannual Sales: March and September. *Discount to Nonprofit Organizations:* 10 percent. *Repairs:* Skis. *Rent:* Tents, stoves, frame packs, snowshoes, downhill and cross-country skis, ice axes, crampons. *Rental Sale:* March.

The Summit of Twin Sisters, Rocky Mountain National Park
(Photo from the collection of North Country Mountaineering, Inc.)

J. RICH SPORTS LTD.   303-476-0860
1000 Lions Ridge Loop
P.O. Box 3164
Vail, Colorado 81657
   Rent: Downhill and cross-country skis.

WHEAT RIDGE

A-1 RENTAL   303-424-4457
7080 W. 38 Avenue
Wheat Ridge, Colorado 80033
   Rent: Sleeping bags, foam pads, tents, stoves, frame packs, soft
packs, snowshoes.

HAMDEN

SKI HUT INCORPORATED   203-248-4475
2840 Whitney Avenue
Hamden, Connecticut 06518
   Discount to Nonprofit Organizations: Sometimes—it's worth asking. *Repairs:* Minor repairs on packs and bags. *Rent:* Tents, frame packs, snowshoes, downhill skis, kayaks, cross-country skis. *Rental Sale:* October. *R.P.*

HARTFORD

EASTERN MOUNTAIN SPORTS   203-278-7105
1 Civic Center Plaza
Hartford, Connecticut 06103
   Semiannual Sales: April and October. *Discount:* Their own brand represents a saving of about 10 to 20 percent over comparable name brand gear. *Discount to Nonprofit Organizations:* Only on bulk orders, 10 to 15 percent. *Manufacturers' Overruns and Seconds:* 10 to 30 percent off list price. *Rent:* Sleeping bags, foam pads, tents, frame packs, soft packs, snowshoes, cross-country skis, ice axes. *R.P.*

LITCHFIELD

WILDERNESS SHOP   203-567-5905
Sports Village, Route 202
Litchfield, Connecticut 06759
   Discount to Nonprofit Organizations: 10 to 15 per cent. *Clearance Center:* 40 percent off on brands that may at times include Raichle, Kelty, North Face, Coleman, Jansport. *Repairs:* Cross-country ski repairs. *Rent:* Sleeping bags, foam pads, tents, stoves, frame packs, snowshoes, canoes, cross-country skis. *R.P.*

ORANGE

SURPLUS TRADING POST INCORPORATED   203-799-2037
153 Boston Post Road
Orange, Connecticut 06477
   Discount: They sell some merchandise at less than manufacturers' suggested list price. *Discount to Nonprofit Organizations:* They do give them. Check with the store. *Repairs:* Scuba tanks and regulators.

WEST HARTFORD

CLAPP & TREAT, INCORPORATED   203-236-0878
674 Farmington Avenue
West Hartford, Connecticut 06119
   Discount to Nonprofit Organizations: Depends upon quantity; varies from 10 to 20 percent. *Repairs:* Stoves, lanterns, minor repairs on boots. *Rent:* Tents, stoves, frame packs, snowshoes, cross-country skis. *Rental Sale:* First week in March. *R.P.*

WEST SIMSBURY

GREAT WORLD   203-658-4461
250 Farms Village Road
Box 250
West Simsbury, Connecticut 06042
   Annual Sale: Check with them for the date of this year's sale. *Discount to Nonprofit Organizations:* 10 percent. *Clearance Center:* 30 to 40 percent off brands that may at times include Vasque Voyageur, Wilderness Experience, and Alpine Designs. *Repairs:* Canoes, kayaks, cross-country skis. *Rent:* Sleeping bags, foam pads, tents, frame packs, snowshoes, kayaks, canoes, cross-country skis. *Rental Sale:* Each fall. *R.P.* Only for skis.

WILTON

SKI HUT INCORPORATED   203-762-8324
Keeler Building
Wilton, Connecticut 06897
   Repairs: Minor repairs. *Rent:* Tents, frame packs, snowshoes,

downhill and cross-country skis, kayaks. *Rental Sale:* October. *R.P.*

Delaware

DOVER

DOVER ARMY-NAVY STORE   302-736-1959
222 Loockerman Street
Dover, Delaware 19901
   Manufacturers' Seconds: 10 to 15 percent off list price. *Discount to Nonprofit Organizations:* 10 percent.

NEW CASTLE

CAMPERS CORNER   302-328-8975
500 School Lane
New Castle, Delaware 19720
   Semiannual Sales: March and November. *Discount:* 10 to 30 percent off list price on brands like Dunham and Camp Trails. *Repairs:* Coleman-Primus.

NEWARK

WICK'S SKI SHOP   302-737-2521
Chestnut Hill Plaza
Newark, Delaware 19713
   Repairs: Ski equipment. *Rent:* Downhill and cross-country skis, kayaks, canoes.

WILMINGTON

WICK'S SKI SHOPS   302-798-1818
1201 Philadelphia Pike
Wilmington, Delaware 19809
   Rent: Downhill and cross-country skis, kayaks, canoes.

WILDERNESS CANOE TRIPS   302-658-0515
1002 Parkside Drive
Oak Lane Manor
Wilmington, Delaware 19803
   Discount to Nonprofit Organizations: They do give it. Ask. *Repairs:* Minor repairs. *Rent:* Tents, stoves, frame packs, kayaks, canoes. *R.P.*

────────────────────────────── **District of Columbia**

EDDIE BAUER, INC.   202-331-8009
Corner of Q and M Streets
Washington, D.C. 20007
   Semiannual Sales: Pre-Christmas and the first week of June.

HERMAN'S WORLD OF SPORTING GOODS   202-638-6434
800 E Street, N.W.
Washington, D.C. 20004
   Annual Sale: Around November 28 each year—also, watch your newspaper for additional sales. *Discount:* 5 to 10 percent on some merchandise, especially Coleman products.

────────────────────────────────────── **Florida**

ARCADIA

CANOE OUTPOST, INCORPORATED   813-494-1215
Route 2, Box 301
Arcadia, Florida 33821
   Discount to Nonprofit Organizations: 20 percent. *Rent:* Sleeping bags, foam pads, tents, stoves, canoes.

BRANFORD

CANOE OUTPOST   904-935-1226
Box 473
Branford, Florida 32008
   Discount to Nonprofit Organizations: 20 percent. *Rent:* Sleeping
bags, foam pads, tents, stoves, canoes.

COCOA

THE WILDERNESS SHOP   305-632-3070
1426 Lake Drive—P.O. Box 3325
Cocoa, Florida 32922
   Discount to Nonprofit Organizations: 10 percent. *Repairs:* Optimus stoves and Coleman appliances.

GAINESVILLE

TRAIL SHOP   904-372-0521
1518 North West 13 Street
Gainesville, Florida 32601
   Discount: They occasionally sell at less than list price, but it's not
a regular practice. *Discount to Nonprofit Organizations:* Occasionally—10 percent. *Rent:* Frame packs.

JACKSONVILLE

TRAIL SHOP   904-744-2292
1421 University Boulevard North
Jacksonville, Florida 32211
   Discount to Nonprofit Organizations: 10 percent. *Rent:* Frame
packs.

MIAMI

TROPICAL WILDERNESS OUTFITTERS   305-253-8131
15355 South Dixie Highway
Miami, Florida 33157
   Discount to Nonprofit Organizations: 10 to 20 percent. *Repairs:* Tents, sleeping bags, packs. *Rent:* Sleeping bags, foam pads,

parkas, tents, stoves, frame packs, soft packs, kayaks, canoes, crampons, ice axes; they will also provide the use of snowshoes, cross-country skis and downhill skis but only for expeditions. *Rental Sale:* They hold an annual auction. Ask about this year's date. *R.P. Special Features:* They teach courses in tropical survival.

NOBLETON

CANOE OUTPOST   904-796-4343
Box 188
Nobleton, Florida 33554
   Discount to Nonprofit Organizations: 20 percent. *Rent:* Sleeping bags, foam pads, tents, stoves, canoes.

SAINT PETERSBURG

CAMPER'S GEAR   813-822-7592
1935 1st Avenue South
Saint Petersburg, Florida 33712
   Discount to Nonprofit Organizations: 10 percent. *Manufacturers' Overruns:* They're the only source we know of in Florida for manufacturers' overruns—10 to 40 percent discount. They carry a very respectable line of moderate price gear, so do check them out. *Repairs:* Stoves, lanterns. *Rent:* Tents, stoves, frame packs, canoes. *Rental Sale:* They sell off their canoes after about eleven rentals. Ask if they have any they want to sell off.

TALLAHASSEE

TRAIL SHOP   904-222-5608
206 West College Avenue
Tallahassee, Florida 32301
   Discount to Nonprofit Organizations: 10 percent. *Rent:* Frame packs.

VALRICO

CANOE OUTPOST   813-681-2666
Route 1, Box 414 K
Valrico, Florida 33594

Discount to Nonprofit Organizations: 20 percent. *Rent:* Sleeping bags, foam pads, tents, stoves, canoes.

ATHENS

BAIRS SKI & HIKE SHOP   404-549-2205
825 Baxter Street
Athens, Georgia 30601
　Manufacturers' Seconds: 25 percent off list price. *Repairs:* Ski repairs only. *Rent:* Sleeping bags, foam pads, tents, stoves, frame packs, downhill skis. *R.P.:* Only on demonstration skis.

ATLANTA

BAIRS SKI & HIKE SHOP   404-261-8978
3228 Roswell Road N.E.
Atlanta, Georgia 30305
　Semiannual Sales: March and Labor Day. *Manufacturers' Seconds:* 25 percent off list price. *Repairs:* Ski repairs only. *Rent:* Sleeping bags, foam pads, tents, stoves, frame packs, downhill skis. *R.P.:* Only on demonstration skis.

BLUE RIDGE MOUNTAIN SPORTS LTD.   404-266-8372
Lennox Square
3393 Peach Tree Road N.E.
Atlanta, Georgia 30326
　Semiannual Sales: January and June or July. *Discount to Nonprofit Organizations:* 10 percent. *Clearance Center:* 30 percent off regular price.

GEORGIA OUTDOORS   404-256-4040
6518 Roswell Road
Atlanta, Georgia 30328
　Semiannual Sales: January and June. *Discount to Nonprofit Organizations:* 5 percent. *Manufacturers' Overruns and Seconds:* 20 percent off list price; they carry a very nice line of gear, so watch to see what overruns and/or seconds they get in. *Repairs:* Optimus, Jansport, minor repairs on all products they sell.

HIGH COUNTRY INC.   404-255-4684
6300 Powers Ferry Road N.W.
Atlanta, Georgia 30339

Annual Sales: Change each year. Check with them. *Discount to Nonprofit Organizations:* 5 percent discount. *Manufacturers' Overruns and Seconds:* 15 to 20 percent off list price. *Repairs:* Kayaks, canoes. *Rent:* Sleeping bags, foam pads, tents, stoves, frame packs, kayaks, canoes. *Rental Sales:* Twice each year—ask them for this year's schedule. *R.P.*

ROCKY MOUNTAIN SPORTS   404-252-3157
6125 Roswell Road
Atlanta, Georgia 30328

Semiannual Sales: Labor Day and March 1. *Discount:* According to them, they do and they don't. Apparently they do but are wary about saying so. Stop by and check out their prices. *Discount to Nonprofit Organizations:* 10 percent off their regular prices. *Manufacturers' Overruns:* 20 to 30 percent off list prices. They carry some very fine brands, like Class 5, Ascente, Nordica, and Donner, so it's worth checking to see whose overruns they were able to get. *Repairs:* Ski bindings and boots. *Rent:* Tents, frame packs, soft packs, downhill skis, canoes. *Rental Sale:* They sell their ski equipment after each summer. *R.P.*

DECATUR

GEORGIA OUTDOORS   404-284-5337
1945 Candler Road
Decatur, Georgia 30032

Semiannual Sales: January and June. *Discount to Nonprofit Organizations:* 5 percent. *Manufacturers' Overruns and Seconds:* 20 percent off list price. They carry a very nice line of gear, so watch to see what overruns and/or seconds they get in. *Repairs:* Optimus; Jansport, minor repairs on all products sold. *Rent:* Sleeping bags, tents, frame packs. *Rental Sale:* Mid-June. *R.P.*

HONOLULU

OMAR THE TENT MAN   801-841-0257
1336 A-4 Dillingham Boulevard
Honolulu, Hawaii 96817
   Discount: They sell equipment at less than manufacturers' list price, but didn't specify the size of the discount. They carry a respectable line of equipment: Adventure 16, Kelty, Trailwise, Coleman, Sierra Designs, so it's worth looking into. *Manufacturers' Overruns and Seconds:* 10 to 40 percent off list price. *Repairs:* Hang gliders, trailers, tents, lanterns, stoves; they'll repair any equipment they sell. *Rent:* Sleeping bags, foam pads, parkas, tents, stoves, frame packs, soft packs. *R.P.*

BOISE

KOPPEL'S BROWZEVILLE   208-344-3539
30 Fairview—Box 1174
Boise, Idaho 83701
   Discount: They sell good, reasonably priced gear (brands like White Stag, Coleman, Camp Ways) at 10 to 20 percent below list price. *Discount to Nonprofit Organizations:* 10 percent. *Manufacturers' Overruns and Seconds:* 30 to 40 percent discount.

OLD BOISE BOOTWORKS   208-344-3821
515 Main Street
Boise, Idaho 83702
   Discount to Nonprofit Organizations: 10 to 20 percent. *Repairs:* Sewing repairs; they'll also do custom modification of your pack or sleeping bag. *Rent:* Sleeping bags, tents, hiking boots, climbing boots, stoves, frame packs, soft packs, snowshoes, ice axes, crampons, cross-country skis. *Rental Sales:* Negotiable. *R.P.*

SAWTOOTH MOUNTAINEERING   208-376-3731
5200 Fairview Avenue
Boise, Idaho 83704

Semiannual Sales: September and March. *Repairs:* Clothing, sleeping bags, stoves, tents, cross-country ski equipment, packs—an excellent repair center. *Rent:* Sleeping bags, foam pads, tents, frame packs, soft packs, snowshoes, ice axes, crampons, cross-country skis. *Rental Sales:* September and March. *R.P. Special Features:* Sawtooth Mountaineering is an excellent source of technical mountaineering equipment (at list price).

DRIGGS

MOUNTAINEERING OUTFITTERS, INC.   208-354-2222
Box 116/62 North Main
Driggs, Idaho 83422

Discount: 5 percent below list price on brands like Camp Trails, Peter Storm, Great Pacific Iron Works, Buck, Powderhorn, Paul Petzoldt Wilderness Equipment, MSR, and Rivendell; this is one of *very* few places we've seen where you can get any discount on brands like Rivendell, MSR., and Great Pacific Iron Works. *Discount to Nonprofit Organizations:* 5 percent. *Manufacturers' Overruns and Seconds:* 5 to 50 percent discount on top line gear. *Clearance Center:* 10 to 40 percent off on top brands; as with any clearance center, what they have reduced depends upon what they bought and didn't sell fast enough, or what got slightly soiled or damaged, but any place that offers the possibility of discount MSR stove parts, or any of Rivendell's superb gear is worth watching closely. *Rent:* Cross-country skis. *Rental Sale:* April. *R.P.*

IDAHO FALLS

SOLITUDE SPORTS
475 A Street
Idaho Falls, Idaho 83401

Discount: They occasionally sell gear at less than manufacturers' list price. Discount is 5 to 10 percent. *Discount to Nonprofit Organizations:* 10 percent. *Manufacturers' Seconds:* 10 to 20 percent discount. *Repairs:* Stoves, skis, anything they sell. *Rent:* Sleeping bags, foam pads, tents, frame packs, snowshoes, cross-country skis. *Rental Sale:* End of winter (cross-country skis only). *R.P.*

KETCHUM

THE ELEPHANT'S PERCH   208-726-3497
220 East Avenue
Ketchum, Idaho 83340

Semiannual Sales: September and March. *Discount to Nonprofit Organizations:* 10 to 15 percent. *Manufacturers' Seconds:* They occasionally carry them, at from 20 to 30 percent off list price. The Elephant's Perch carries good equipment generally (Kelty, North Face, Vasque Voyageur) so it pays to check to see what seconds they've been able to get. *Repairs:* Cross-country skis, bicycles. *Rent:* Sleeping bags, foam pads, tents, frame packs, soft packs, snowshoes, cross-country skis. *R.P.*

LEWISTON

FOLLETT'S MOUNTAIN SPORTS   208-743-4200
3206 5th Street
Lewiston, Idaho 83501

Annual Sale: March. *Discount:* Some of their gear is sold at discount—up to 20 percent. *Discount to Nonprofit Organizations:* 10 to 40 percent. *Repairs:* Skis, backpacking equipment, bicycles. *Rent:* Downhill and cross-country skis. *Rental Sale:* Each spring—ask them for this year's dates. *R.P.*

MOSCOW

NORTHWESTERN MOUNTAIN SPORTS   208-882-0133
410 West 3rd
Moscow, Idaho 83843

Semiannual Sales: March or February, and July. *Discount to Nonprofit Organizations:* 10 percent. *Clearance Center:* 50 percent discount; they carry a particularly wide range of boot brands, so you might do well to keep checking here if you're in the market for hiking or light mountaineering boots. *Repairs:* Stoves, downhill skis, tennis rackets. *Rent:* Downhill and cross-country skis. *Rental Sales:* March and September. *R.P.*

NORTHWEST RIVER SUPPLIES, INC.   208-882-2382
214 North Main
Moscow, Idaho 83843
   Semiannual Sales: August and September. *Discount:* 5 to 15
percent off manufacturers' suggested list price on kayak and canoe
equipment. *Discount to Nonprofit Organizations:* 5 to 15 percent.

SUN VALLEY

SNUG MOUNTAINEERING   208-622-9305
Box 598
Sun Valley, Idaho 83353
   Semiannual Sales: Each spring and fall—check with them for the
exact date. *Manufacturers' Overruns and Seconds:* They carry them
sometimes at from 30 to 40 percent off list price. *Rent:* Sleeping
bags, foam pads, tents, frame packs, soft packs, downhill and cross-
country skis, kayaks, bicycles, ice axes, crampons. *R.P.*

TWIN FALLS

NEWTON'S SPORTS CENTER   208-733-8371
1188 Blue Lakes North
Twin Falls, Idaho 83301
   Discount to Nonprofit Organizations: 15 percent. *Repairs:* Minor
repairs.

————————————————————————————————— **Illinois**

CHAMPAIGN

BUSHWACKER, LTD.   217-359-3353
702 South Neil
Champaign, Illinois 61820
   Semiannual Sales: October and February or March. *Discount to
Nonprofit Organizations:* 10 to 15 percent. *Manufacturers' Overruns and Seconds:* 10 to 15 percent off list price; they are worth
looking into as their regular stock consists of some excellent brands
like Fabiano, Jansport, Sierra Designs, and North Face. October and

February sales are the times to look for these. *Repairs:* Minor repairs to soft goods, also limited repairs of cross-country skis, equipment. *Rent:* Sleeping bags, tents, kayaks, cross-country skis. *R.P.*

CHICAGO

CAMPING UNLIMITED   312-286-1414
5207 North Milwaukee Avenue
Chicago, Illinois 60630
 Annual Sale: End of February. *Discount:* They do sell some merchandise at less than full list price—worth stopping in for a look. *Rent:* Tents, frame packs, snowshoes, cross-country skis. *Rental Sales:* End of each season. *R.P.:* Partial. *Special Features:* An extremely large stock of freeze-dried and dehydrated food.

EREWHON MOUNTAIN SUPPLY   312-262-3832
1252 West Devon Street
Chicago, Illinois 60660
 Semiannual Sales: February and August. *Discount to Nonprofit Organizations:* 10 to 15 percent. *Rent:* Sleeping bags, tents, foam pads, frame packs, cross-country skis.

TRAVELER'S ABBEY   312-549-3230
2934 North Broadway
Chicago, Illinois 60657
 Annual Sale: February. *Discount to Nonprofit Organizations:* 10 percent. *Repairs:* Skis, packs. *Rent:* Sleeping bags, foam pads, tents, stoves, frame packs, snowshoes, cross-country skis. *Rental Sale:* Each fall—ask for this year's date.

WIN SUM SKI AND MOUNTAIN SHOP, LTD.   312-751-1776
455 West Armitage
Chicago, Illinois 60614
 Semiannual Sales: End of each season—check with them for the exact dates. *Manufacturers' Overruns and Seconds:* 30 percent discount—should be interesting; Win Sum carries a superb line of gear. *Repairs:* Skis. *Rent:* Sleeping bags, tents, frame packs, soft packs, downhill and cross-country skis. *Rental Sales:* End of each season. *R.P.*

DES PLAINES

EREWHON MOUNTAIN SUPPLY   312-298-1840
1522 Miner Street
Des Plaines, Illinois 60016
   Semiannual Sales: February and August. *Discount to Nonprofit Organizations:* 10 to 15 percent. *Manufacturers' Seconds:* 10 percent discount off list price. *Rent:* Sleeping bags, foam pads, tents, frame packs, snowshoes, cross-country skis.

MAYWOOD

EASY CAMPING   321-344-4454
510 South 5th Avenue
Maywood, Illinois 60153
   Semiannual Sales: Spring and fall—check with them for dates of this year's sales. *Discount to Nonprofit Organizations:* 10 percent. *Repairs:* Cross-country ski equipment. *Rent:* Cross-country skis.

PEORIA

BUSHWACKER LTD.   309-676-1038
1241 West Main Street
Peoria, Illinois 61606
   Semiannual Sales: October and February or March. *Discount to Nonprofit Organizations:* 10 to 15 percent. *Manufacturers' Seconds:* 10 to 15 percent off list price; they carry top line gear. *Repairs:* Soft goods, cross-country ski equipment. *Rent:* Sleeping bags, foam pads, tents, stoves, frame packs, kayaks, cross-country skis. *R.P.*

SPRINGFIELD

BUSHWACKER LTD.   217-787-7692
948 Clocktower Drive   217-787-7685
Springfield, Illinois 62704
   Semiannual Sales: October and February or March. *Discount to Nonprofit Organizations:* 10 to 15 percent. *Manufacturers' Seconds:* 10 to 15 percent off list price; they carry top line gear.

Repairs: Soft goods, cross-country ski equipment. *Rent:* Sleeping bags, foam pads, tents, stoves, frame packs, kayaks, cross-country skis. *R.P.*

WOODSTOCK

OUTDOOR RECREATION, INC.   815-338-6088
1801½ South Route 47
Woodstock, Illinois 60098
 Discount to Nonprofit Organizations: 10 to 15 percent. *Clearance Center:* 20 to 25 percent discount on brands that may include Vasque Voyageur, North Face, Snow Lion, Wilderness Experience, and Edge boots. *Repairs:* Minor repairs. *Rent:* Sleeping bags, foam pads, tents, stoves, frame packs, snowshoes, canoes, ice axes, crampons, cross-country skis. *Rental Sale:* Each fall—ask them for this year's date. *R.P.*

─────────────────────────────── **Indiana**

BLOOMINGTON

GREEN MOUNTAIN SUPPLY   812-334-1845
322 East Kirkwood
Bloomington, Indiana 47401
 Discount to Nonprofit Organizations: 20 percent. *Clearance Center:* 10 to 20 percent off name brands that may at times include Class 5, Camp 7, Fabiano, Vasque, and North Face. *Repairs:* Stoves and stitching. *Rent:* Stoves, cross-country skis, kayaks, canoes. *R.P. Special Features:* They carry a nice line of caving equipment, some of which they'll rent.

EVANSVILLE

PINE MOUNTAIN EQUIPMENT   812-476-2684
1529 S. Greenriver Road
Evansville, Indiana 47715
 Rent: Sleeping bags, tents, foam pads, stoves, frame packs, soft packs, canoes. *R.P.*

FORT WAYNE

ROOT'S CAMP 'N SKI HAUS   219-484-2604
6844 North Clinton Street
Fort Wayne, Indiana 46825
  Discount to Nonprofit Organizations: 10 percent, but on large
purchases only. *Rent:* Sleeping bags, tents, foam pads, downhill and
cross-country skis, canoes. *Rental Sale:* Late fall. *R.P.*

WATER MEISTER SPORTS   219-432-0011
P.O. 5026
3211 Covington Road
Fort Wayne, Indiana 46805
  Discount to Nonprofit Organizations: They give one, but didn't
specify how much. *Rent:* Sleeping bags, foam pads, tents, stoves,
frame packs, snowshoes, kayaks, canoes, cross-country skis. *R.P.*
Special Features: They produce a wide range of custom equipment
including flotation bags, paddles, kayaks, and spray skirts. They also
offer instruction in fencing and caving.

PITTSBORO

INDIANA CAMP SUPPLY, INC.   317-547-1134
P.O. Box 344
Pittsboro, Indiana 46167
  Discount: All fairly large food orders are discounted: 10 percent
on orders of $100 or more, 25 percent on orders of $250 or more.
This is a terrific idea; you'll probably be using more than $50 worth
of food over a year anyway. If not, get together with a friend, or a
group of friends, and get this really fine discount. Buying with four
friends ought to qualify you for the 25 percent discount. If you are
buying food for a larger group (i.e., an outing club), you may be able
to negotiate a larger discount—or you may want to look into the
possibility of buying it wholesale; look at Chapter 8 for details on
forming an equipment-buying cooperative. The advantage of buying from Indiana Camp Supply, of course, is that you can mix brands
—and they carry every major brand. Indiana Camp Supply also sells
the rest of their gear at about 10 percent less than manufacturers'
suggested list price. Do check them out, as they carry such fine
brands as Snow Lion and Sierra West. As far as we're concerned,
this is a very hip operation, at least from a discount buyer's point

of view. *Discount to Nonprofit Organizations:* 20 percent. *Special Features:* If you wish, they'll help you with your meal planning.

SOUTH BEND

SIERRA MOUNTAIN SUPPLY   219-233-8383
2216 Miami
South Bend, Indiana 46613
   Discount to Nonprofit Organizations: 20 percent. *Clearance Center:* 10 to 20 percent discount off list price on brands that may at times include Class 5, Camp 7, Fabiano, Vasque, and North Face. *Repairs:* Stoves and stitching. *Rent:* Frame packs, stoves, cross-country skis, kayaks, canoes. *R.P. Special Features:* They carry a nice line of caving equipment, some of which they will rent.

Iowa

IOWA CITY

BIVOUAC   319-338-7677
32 South Clinton
Iowa City, Iowa 52240
   Discount: Up to 10 percent discount; they carry brands like Fabiano, Trailwise, Jansport, Snow Lion. *Rent:* Sleeping bags, tents, stoves, frame packs, cross-country skis. *Rental Sales:* After Christmas, and after the summer.

Kansas

LAWRENCE

GRAN SPORTS   913-843-3328
7th and Arkansas
Lawrence, Kansas 66044
   Rent: Tents, stoves, frame packs.

RIDE-ON OUTDOOR SPORTS   913-843-8484
1401 Massachusetts
Lawrence, Kansas 66044

Sales: Ski merchandise is sold mid-February. Tennis goods go on sale during July. Backpacking gear goes on sale in October. *Discount:* 10 to 20 percent on many items. *Manufacturers' Overruns and Seconds:* Up to 25 percent off list price. *Clearance Center:* Up to 25 percent off list price. *Repairs:* Bicycles and boots. *Rent:* Tennis rackets, bicycles, mopeds. *Sale:* End of season. *R.P.*

OVERLAND PARK

RIDE-ON OUTDOOR SPORTS   913-492-9494
Oak Park Mall
Overland Park, Kansas 66210

Sales: Ski merchandise is sold mid-February. Tennis goods go on sale during July. Backpacking gear goes on sale in October. *Discount:* 10 to 20 percent on many items. *Manufacturers' Overruns and Seconds:* Up to 25 percent off list price. *Clearance Center:* Up to 25 percent off list price. *Repairs:* Bicycles and boots. *Rent:* Tennis rackets, bicycles, mopeds.

SHAWNEE MISSION

VOYAGEUR'S PACK & PORTAGE SHOP   913-262-6611
5935 Merriam Drive
Shawnee Mission, Kansas 66203

Annual Sales: January or February—check with them. *Discount to Nonprofit Organizations:* 10 percent. *Manufacturers' Overruns and Seconds:* 20 to 30 percent when they have them—occasionally. *Repairs:* Hang gliders. *Rent:* Sleeping bags, foam pads, tents, stoves, frame packs, soft packs, kayaks, canoes, cross-country skis. *Rental Sale:* January or February. *R.P.*

TOPEKA

GRAN SPORT   913-354-1656
527 Topeka Avenue
Topeka, Kansas 66603

Discount to Nonprofit Organizations: They give one but didn't tell us how much. Check with them.

WICHITA

BACKWOODS EQUIPMENT CO.   316-751-7376
144 North Market
Wichita, Kansas 67202
  Discount to Nonprofit Organizations: Varies with size of ordeʀ. *Manufacturers' Overruns and Seconds:* Good brands at 20 percent discount. *Rent:* Sleeping bags, foam pads, tents, frame packs, soft packs, kayaks, cross-country skis. *Rental Sale:* February. *R.P.*

MOUNTAIN HIGH, INC.   316-684-6579
2936 East Douglas
Wichita, Kansas 67214
  Semiannual Sales: Late summer, late winter. *Discount to Nonprofit Organizations:* 10 percent. *Repairs:* Minor stove, pack, and ski repairs. *Rent:* Sleeping bags, foam pads, tents, frame packs, snowshoes, ice axes, cross-country skis. *Rental Sales:* September and March. *R.P.*

Kentucky

LEXINGTON

**SAGE, SCHOOL OF THE OUTDOORS,
OUTFITTERS, EXPEDITIONERS**   606-255-1547
209 East High Street
Lexington, Kentucky 40507
  Semiannual Sales: Spring and fall—check with them. *Discount to Nonprofit Organizations:* 5 percent, to Boy Scouts only. *Repairs:* Stoves, carbide lamps. *Rent:* Sleeping bags, foam pads, tents, stoves, frame packs, kayaks, canoes.

LOUISVILLE

THE GREAT OUTDOORS   502-895-7353
3824 Wilmington Avenue
Louisville, Kentucky 40207
  Sales: They don't have a regular sale schedule. Check with them about impending special sales. *Rent:* Stoves, frame packs, soft

packs, cross-country skis. *Rental Sales:* Negotiable—ask them. *R.P.*
Special Features: They give hang gliding lessons.

SPORT CENTER   502-583-1376
628 South 4th Street
Louisville, Kentucky 40202
   Discount to Nonprofit Organizations: It varies. They carry military
surplus and moderately priced camping supplies, so this might be
a good place to shop for younger Scout groups.

VIKING CANOE & MOUNTAINEERING
CENTER   502-361-8051
3308 Preston Highway
Louisville, Kentucky 40213
   Clearance Center: They didn't tell us what the savings would be,
but brands you might find there can occasionally include Jansport,
Raichle, Fabiano, Galibier, Alpine Designs, and Ascente. You might
as well watch the clearance center; Louisville is rough going for
backpacking gear bargain hunters. *Rent:* Sleeping bags, foam pads,
tents, stoves, frame packs, soft packs, kayaks, canoes. *R.P.*

_____ **Louisiana**

BATON ROUGE

THE BACKPACKER   504-383-6670
3378 Highland Road
Baton Rouge, Louisiana 70802
   Repairs: Canoes and tents. *Rent:* Sleeping bags, tents, stoves,
frame packs, kayaks, canoes. *R.P. Special Features:* They offer a
discount on rentals to nonprofit organizations.

LAFAYETTE

PACK AND PADDLE, INC.   318-232-5854
601 Pinhook Road East
Lafayette, Louisiana 70501
   Discount to Nonprofit Organizations: 10 percent discount on
purchases over $400. *Rent:* Sleeping bags, foam pads, parkas,
tents, frame packs, kayaks, canoes. *Rental Sales:* October and
November.

NEW ORLEANS

CANOE & TRAIL SHOP   504-488-8528
624 Moss Street
New Orleans, Louisiana 70119
   Discount to Nonprofit Organizations: 10 percent. *Repairs:* Canoes, packs, stoves. *Rent:* Sleeping bags, foam pads, parkas, tents, stoves, frame packs, kayaks, canoes. *Rental Sale:* November.

GUS BETAT'S COUNTRY ROADS, INC.   504-861-0008
8623 Oak Street
New Orleans, Louisiana 70118
   Discount to Nonprofit Organizations: 10 percent. *Repairs:* Canoes. *Rent:* Sleeping bags, tents, stoves, frame packs, soft packs, canoes. *R.P.*

RIVER RIDGE

LADS, INC.   504-737-3881
9031 Jefferson Highway
River Ridge, Louisiana 70123
   Discount to Nonprofit Organizations: 15 percent—the best we've found in this state. *Repairs:* All repairs. *Rent:* Tents, sleeping bags, stoves, camp cots, Coleman lamps, water skis. *R.P.*

SHREVEPORT

KAMPER'S KORNER   318-686-7116
3435 West 70th Street
Shreveport, Louisiana 71108
   Discount to Nonprofit Organizations: 10 percent. *Repairs:* All camping equipment. *Rent:* Tents, stoves, soft packs, canoes. *R.P.*

WEST MONROE

OUTDOOR ADVENTURES, INC.   318-387-0128
410 North 6th Street
West Monroe, Louisiana 71291
   Discount: 2 percent below manufacturers' list price—it's better

than nothing. *Discount to Nonprofit Organizations:* 10 percent. *Repairs:* All equipment repairs. *Rent:* Sleeping bags, foam pads, tents, stoves, frame packs, soft packs, canoes. *Rental Sales:* End of season—ask them for this year's date. *R.P.*

—————————————————————————**Maine**

ANDOVER

AKERS SKI   207-392-4582
Andover, Maine 04216
   Discount to Nonprofit Organizations: It varies. *Clearance Center:* This fine mail order firm offers a merchandise clearance sheet for its mail order catalog recipients. *Rent:* Cross-country skis. *R.P.*

BAR HARBOR

NORUMBEGA MOUNTAIN SHOP   207-288-3700
185 Main Street
Bar Harbor, Maine 04609
   Semiannual Sales: August and February. *Discount to Nonprofit Organizations:* 10 to 20 percent. *Rent:* Tents, snowshoes, cross-country skis. *Rental Sales:* September and March. *R.P.*

FARMINGTON

NORTHERN LIGHTS, LTD.   207-778-6566
RFD #2—Routes 2 & 4
Farmington, Maine 04938
   Discount to Nonprofit Organizations: 10 to 20 percent. *Manufacturers' Overruns and Seconds:* 5 to 20 percent off regular list price. *Clearance Center:* 15 to 30 percent. *Repairs:* Bicycles, skis, boats. *Rent:* Snowshoes, downhill skis, canoes, bicycles, cross-country skis. *R.P. Special Features:* They sell woodburning stoves.

FREEPORT

L. L. BEAN, INC.   207-865-3111
Freeport, Maine 04032
  Rent: Canoes, cross-country skis—open 24 hours. *R.P.*

FRYEBURG

SACO BOUND/NORTHERNWATERS   603-447-2177
Fryeburg, Maine 04037
  Discount: They sell some merchandise at less than list price.
Discount to Nonprofit Organizations: 15 percent discount for five
or more canoes. *Manufacturers' Seconds:* They do occasionally
have them, but didn't tell us at what percentage of savings.
Rent: Sleeping bags, tents, stoves, soft packs, kayaks, canoes.
Rental Sales: At any time. *R.P.*

────────────────────────────────────── **Maryland**

ANNAPOLIS

SUMMIT MOUNTAINEERING   301-268-8805
108 Old Solomon Island Road
Annapolis, Maryland 21401
  Semiannual Sales: Fall and spring—check with them for exact
dates. *Discount to Nonprofit Organizations:* 10 percent. *Manufac-
turers' Overruns and Seconds:* 15 percent off regular list price.
Repairs: Stoves. *Rent:* Sleeping bags, tents, stoves, frame packs.
Rental Sales: Fall and spring—ask them for this year's dates. *R.P.*
Special Features: They custom construct tent bags, stuff sacks, rain
parkas, mitten shells, booties.

BALTIMORE

H & H CAMPER'S HAVEN   301-752-2580
424 North Eutaw Street
Baltimore, Maryland 20201
  Rent: Tents, stoves, frame packs, canoes.

CUMBERLAND

TAYLOR RENTAL CENTER   301-777-5001
39 Queen Street
Cumberland, Maryland 21502
   Rent: Sleeping bags, tents, stoves, canoes, bicycles, Coleman lanterns, coolers, cots, bicycle racks, ski racks.

ELLICOTT CITY

APPALACHIAN OUTFITTERS   301-465-7227
8563 Route 40 West (Pike)
Ellicott City, Maryland 21043
   Discount: They generally sell gear at a 5 to 10 percent discount. Brands they carry include North Face, Gerry, Jansport, Alpenlite, Camp 7. *Repairs:* Stoves. *Rent:* Sleeping bags, foam pads, stoves, frame packs, shoes, cross-country skis. *R.P.*

GAITHERSBURG

HUDSON BAY OUTFITTERS·   301-948-2474
315 East Diamond Avenue
Gaithersburg, Maryland 20760
   Discount: Some items sold at 20 to 25 percent off list price. *Discount to Nonprofit Organizations:* Excellent discount: 15 to 25 percent. *Clearance Center:* 50 percent savings on brands that may include Hine/Snowbridge, Kelty, Trailwise, Camp Trails, and Coleman. *Rent:* Sleeping bags, foam pads, tents, stoves, frame packs, soft packs, snowshoes, kayaks, canoes, cross-country skis. *R.P.*

————————————————————————— **Massachusetts**

AMHERST

EASTERN MOUNTAIN SPORTS   413-253-9504
Route 9, Amherst-Hadley Line
Mailing Address: Box 12
Amherst, Massachusetts 01002

Discount: Their own brand of gear sells for about 10 to 20 percent less than comparable name brand merchandise. *Rent:* Sleeping bags, foam pads, Mickey Mouse boots, tents, stoves, frame packs, soft packs, snowshoes, cross-country skis. *Rental Sales:* Spring and fall. *R.P.*

ANDOVER

MOOR & MOUNTAIN   617-475-3665
63 Park Street
Andover, Massachusetts 01810
   Annual Sale: Each spring—check with them for exact date. *Discount to Nonprofit Organizations:* 15 to 25 percent, depending upon volume. *Manufacturers' Seconds:* They carry seconds of Old Town and Mad River canoes at 10 to 15 percent discount. *Repairs:* Fiberglass kayaks, sleeping bags, zippers. *Rent:* Snowshoes, kayaks, canoes, cross-country skis. *R.P.* Sometimes. *Special Features:* They custom construct overboots and gaiters.

BELMONT

CANOE ADVENTURES NO'EAST
(formerly Canoe Camp Outfitters)   617-484-6571
8 Cherry Street
Belmont, Massachusetts 02178
   Discount: They sell brands like Camp Trails and Gerry at 20 percent below manufacturers' suggested list price. *Discount to Nonprofit Organizations:* 25 percent. *Manufacturers' Overruns and Seconds:* 30 percent below list price. *Repairs:* Aluminum canoes. *Rent:* Sleeping bags, packs, tents, frame packs, soft packs, canoes, canoe ditty bags and accessories, Iliad white water paddles, Omega PFD's, Dutch ovens, car top racks, cooking equipment for large groups. *R.P. Special Features:* They will custom construct reflector ovens.

BOSTON

EASTERN MOUNTAIN SPORTS   617-482-4414
One Winthrop Square
Boston, Massachusetts 02108
   Semiannual Sales: October and end of April. *Discount:* Their own

brand of gear generally represents a saving of 10 to 20 percent off the price of comparable name brand gear. *Discount to Nonprofit Organizations:* On large purchases only. *Manufacturers' Overruns and Seconds:* 15 to 30 percent off list price.

EASTERN MOUNTAIN SPORTS   617-254-4250
1041 Commonwealth Avenue
Boston, Massachusetts 02215

Semiannual Sales: October and end of April. *Discount:* Their own brand of gear generally represents a saving of 10 to 20 percent off the price of comparable name brand gear. *Manufacturers' Overruns and Seconds:* 15 to 30 percent off list price. *Clearance Center:* Their Bargain Basement is one of the best clearance centers in the United States: anywhere from 15 to 50 percent savings on boots, skis, tents, parkas—on name brands like Molitor, Lovett skis, Kelty packs, Alpine Designs parkas, Alpine Designs packs, North Face jackets, Eureka tents, Mountain Products tents, etc. *Rent:* Ice axes, crampons, Mickey Mouse boots, tents, foam pads, stoves, frame packs, soft packs, cross-country skis. *Rental Sales:* October and end of April. *R.P.*

BRIGHTON

WILDERNESS HOUSE   617-277-5858
124 Brighton Avenue
Brighton, Massachusetts 02134

Discount: They do discount—usually 5 percent on $250 worth of merchandise, 10 percent on orders over $500. *Discount to Nonprofit Organizations:* 5 percent on orders of $250 or more, 10 percent on orders over $500. *Clearance Center:* They have considerable reductions in their clearance center—from 40 to 80 percent on brands that may at times include Fabiano, Class 5, Mountain-10, as well as their own Mountain Master line. *Rent:* Sleeping bags, foam pads, Mickey Mouse boots, tents, hiking boots, climbing boots, frame packs, snowshoes, canoes, ice axes, cross-country skis. *R.P.*

CAMBRIDGE

BACKPACKER'S COUNTRY   617-868-7464
10 B Mount Auburn Street
Cambridge, Massachusetts 02138

Discount: They discount name brands like Trailwise, Twin Peaks, Camp 7, and Class 5, but the discount varies according to their supply situation—as much as 20 percent off list price. *Manufacturers' Seconds:* Good seconds at about 22 percent below list price. *Repairs:* Soft goods and skis. *Rent:* Sleeping bags, foam pads, Mickey Mouse boots, tents, stoves, frame packs, soft packs, kayaks, canoes, cross-country skis. *Rental Sale:* End of season. *R.P.*

COHASSET

STOW-A-WAY SPORTS   617-383-9116
166 Cushing Highway (Rt. 3A)
Cohasset, Massachusetts 02025
  Rent: Cross-country skis.

CONCORD

MOOR AND MOUNTAIN   617-369-4186
67 Main Street
Concord, Massachusetts 01742
  Annual Sale: Late February or early March. *Discount:* Sometimes, but usually not on name brands; as they wrote us, "It is frowned upon." *Discount to Nonprofit Organizations:* 15 to 20 percent on large quantity orders. *Manufacturers' Seconds:* 10 to 30 percent discount; very occasionally the discount is a bit higher. *Repairs:* Canoes, kayaks, stoves, some skis; they send out clothing, tents, sleeping bags, etc., to their own repair services.

IPSWICH

ALL YEAR ROUND, INC.   617-356-0131
Route 1A
Ipswich, Massachusetts 01938
  Discount to Nonprofit Organizations: 5 percent. *Rent:* Sleeping bags, foam pads, frame packs, soft packs, snowshoes, kayaks, canoes, ice axes, crampons, cross-country skis. *Rental Sale:* End of each season. *R.P.:* Only on skis.

LENOX

ARCADIAN SHOP   413-637-3010
44 Housatonic Street
Lenox, Massachusetts 01240

Discount to Nonprofit Organizations: 10 to 20 percent. *Manufacturers' Overruns:* They have them occasionally. *Repairs:* Cross-country skis. *Rent:* Sleeping bags, foam pads, tents, stoves, frame packs, soft packs, snowshoes, kayaks, cross-country skis. *R.P.:* Skis only.

NORTHAMPTON

DON GLEASON'S CAMPERS SUPPLY, INC.   413-584-4895
9 Pearl Street
Northampton, Massachusetts 01060

Discount: Up to 25 percent discount on name brand gear from such manufacturers as Camp Trails, Coleman, Peak 1, Camp 7, North Face, and Alpine Designs. *Repairs:* Tents.

WELLESLEY

EASTERN MOUNTAIN SPORTS   617-237-2645
189 Linden Street
Wellesley, Massachusetts 02181

Semiannual Sales: October and end of April. *Discount:* Their own brand of gear generally represents a saving of 10 to 20 percent off the price of comparable name brand gear. *Discount to Nonprofit Organizations:* On large orders only.*Manufacturers' Overruns and Seconds:* 15 to 30 percent off list price.

WEST SPRINGFIELD

PADDLERS & PACKERS   413-737-0267
1615 Riverdale Street
West Springfield, Massachusetts 01089

Discount to Nonprofit Organizations: 15 to 20 percent. *Repairs:* Stoves, canoes, other gear. *Rent:* Canoes, cross-country skis.

WILLIAMSTOWN

ARCADIAN SHOP   413-458-3670
1 Water Street
Williamstown, Massachusetts 01267
   Discount to Nonprofit Organizations: 20 percent. *Rent:* Sleeping
bags, foam pads, tents, stoves, frame packs, snowshoes, cross-
country skis. *Rental Sale:* Every two years—ask.

WORCESTER

STRAND'S SKI SHOP INC.   617-852-4333
One West Boylston Drive
Worcester, Massachusetts 01606
   Semiannual Sales: Spring and fall—check with them for this
year's dates. *Discount to Nonprofit Organizations:* 10 to 20 percent. *Manufacturers' Seconds:* 30 to 60 percent discounts, when
they manage to get the merchandise. Do check; they carry some
good brands. *Clearance Center:* 33 to 50 percent reductions off list
price. *Repairs:* They resole boots. *Rent:* Tents, stoves, frame packs,
soft packs, snowshoes, downhill and cross-country skis, ice axes,
crampons. *R.P.*

Michigan

ANN ARBOR

BIVOUAC   313-761-6207
330 South State Street
Ann Arbor, Michigan 48108
   Annual Sale: January. *Rent:* Sleeping bags, tents, stoves, frame
packs, cross-country skis. *Rental Sale:* After Christmas, and after the
summer.

RAUPP CAMPFITTERS   313-769-5574
637 South Main
Ann Arbor, Michigan 48103
   Semiannual Sales: August and February. *Manufacturers' Overruns:* 30 percent discount off list price. *Rent:* Tents, stoves,

frame packs, snowshoes, cross-country skis. *Rental Sale:* Skis in summer. *R.P.*

BIRMINGHAM

THE SPORTSMAN   313-646-1225
184 Pierce Street
Birmingham, Michigan 48011
   Discount: Some items are sold at less than manufacturers' suggested list price. *Discount to Nonprofit Organizations:* It depends upon volume.

FARMINGTON HILLS

THE BENCHMARK   313-477-8116
29450 West Ten Mile Road
Farmington Hills, Michigan 48024
   Repairs: Skis, stoves, packs; minor saving on bags and tents. *Rent:* Frame packs, snowshoes, cross-country skis. *Rental Sale:* End of season. *R.P.*

GRAND RAPIDS

BILL & PAUL'S SPORTHAUS   616-949-0190
Eastbrook Mall
Grand Rapids, Michigan 49508
   Discount to Nonprofit Organizations: 15 to 20 percent. *Manufacturers' Seconds:* 10 to 20 percent discount. *Repairs:* They will resole cross-country ski boots and downhill ski boots. *Rent:* Tents, frame packs, soft packs, kayaks. *Rental Sale:* End of season. *R.P.*

BILL & PAUL'S SPORTHAUS   616-364-0006
North Kent Mall
Grand Rapids, Michigan 49505
   Discount to Nonprofit Organizations: 15 to 20 percent. *Manufacturers' Seconds:* 10 to 20 percent discount. *Repairs:* They will resole cross-country ski boots and downhill ski boots.

KALAMAZOO

RAUPP CAMPFITTERS 616-344-1337
1801 West Main
Kalamazoo, Michigan 49007
 Semiannual Sales: August and February. *Manufacturers' Overruns and Seconds:* 20 percent less than list price. *Repairs:* Stoves. *Rent:* Tents, stoves, frame packs, snowshoes, cross-country skis. *Rental Sales:* Skis in summer. *R.P.*

LANSING

RAUPP CAMPFITTERS 517-484-9401
2208 East Michigan Avenue
Lansing, Michigan 48912
 Semiannual Sales: August and February. *Manufacturers' Overruns and Seconds:* 20 percent discount. *Rent:* Tents, stoves, frame packs, snowshoes, cross-country skis. *Rental Sale:* Skis in summer. *R.P.*

MILFORD

HEAVNER CANOE RENTAL 313-685-2379
2775 Garden Road
Milford, Michigan 48042
 Discount: 5 to 20 percent off list price on some canoes and cross-country skis. *Discount to Nonprofit Organizations:* 10 percent. *Rent:* Kayaks, canoes, cross-country skis. *Rental Sales:* End of seasons.

OAK PARK (DETROIT)

EDDIE BAUER 313-967-0155
21110 Greenfield Road
Oak Park—Detroit, Michigan 48076
 Semiannual Sales: The sale times vary. Check with Richard Fowler, the store's manager.

REDFORD TOWNSHIP

RAUPP CAMPFITTERS   313-534-4044
24000 Plymouth Road
Redford Township, Michigan 48239
   Semiannual Sales: August and February. *Manufacturers' Overruns and Seconds:* 20 percent discount. *Rent:* Tents, stoves, frame packs, snowshoes, cross-country skis. *Rental Sales:* Skis in summer. *R.P.*

ROCHESTER

THE BENCHMARK   313-651-7444
870 S. Rochester Road
Rochester, Michigan 48063
   Rent: Frame packs, snowshoes, cross-country skis. *R.P.*

ROYAL OAK

RAUPP CAMPFITTERS   313-547-6969
421 South Washington
Royal Oak, Michigan 48067
   Semiannual Sales: August and February. *Discount to Nonprofit Organizations:* 10 percent. *Manufacturers' Overruns and Seconds:* 20 percent discount. *Rent:* Tents, stoves, frame packs, snowshoes, cross-country skis. *Rental Sales:* Skis in summer. *R.P.*

Minnesota

DULUTH

CONTINENTAL SKI SHOP   218-728-4466
1305 East 1st Street
Duluth, Minnesota 55805
   Discount to Nonprofit Organizations: 15 percent. *Manufacturers' Overruns and Seconds:* 20 to 25 percent discount. *Repairs:* Skis.

CONTINENTAL SKI SHOP II   218-728-5967
1402 East 1st Street
Duluth, Minnesota 55804
   Discount to Nonprofit Organizations: 15 percent. *Manufacturers' Overruns and Seconds:* 20 to 25 percent discount.

THE SKI HUT   218-724-8525
1032 East 4th Street
Duluth, Minnesota 55805
   Discount to Nonprofit Organizations: 10 to 20 percent. *Manufacturers' Overruns and Seconds:* 10 to 30 percent discount. *Repairs:* Stoves and skis. *Rent:* Cross-country skis. *Rental Sale:* End of season. *R.P.*

ELY

MOOSE LAKE WILDERNESS CANOE TRIPS   218-365-5837
Box 358
Ely, Minnesota 55731
   Annual Sale: May. *Discount to Nonprofit Organizations:* A whopper: 25 percent. *Rent:* Sleeping bags, foam pads, parkas, tents, stoves, canoes, soft packs. *Rental Sale:* Negotiable at any time. *Special Features:* They will custom construct tents.

HIBBING

THE SKI HUT   218-263-6811
Graysher Center
Hibbing, Minnesota 55746
   Semiannual Sales: September and October. *Discount:* They will occasionally sell gear at 10 to 40 percent below list price. *Discount to Nonprofit Organizations:* It varies; check with them. *Repairs:* They repair all the equipment they sell. *Rent:* Kayaks, canoes, cross-country skis, snowshoes, downhill skis. *Rental Sale:* End of season. *R.P.*

MINNEAPOLIS

AARCEE RENTAL CENTER   612-827-5746
2900 Lyndale Avenue, South
Minneapolis, Minnesota 55408
   Annual Sale: Third week in September. *Discount to Nonprofit*

The Woodland Waterfall
(Photo from the collection of North Country Mountaineering, Inc.)

Organizations: 10 percent. *Rent:* Sleeping bags, foam pads, tents, stoves, snowshoes, downhill skis, frame packs, cross-country skis. *Rental Sale:* September. *R.P.:* Applied to new equipment.

EDDIE BAUER, INC.   612-339-9477
821 Marquette
Minneapolis, Minnesota 55402
Semiannual Sales: January or February, and July. *Repairs:* Sleeping bags, down jackets.

FAIR WHEEL BICYCLES & CAMPING GOODS   612-827-4456
2548 Nicollet Avenue South
Minneapolis, Minnesota 55404
Discount: They always give a discount for quantity orders. Also they give as much as a 20 percent discount on a good deal of merchandise. *Discount to Nonprofit Organizations:* As high as 40 percent on some items. *Clearance Center:* 25 to 30 percent discount on brands that may include Trailwise, Mountain Products, Camp Trails, Comfy, Hunter Sleeping Bags, Raichle, and Summit boots. *Rent:* Snowshoes, kayaks, bicycles, cross-country skis. *Rental Sale:* Negotiable at any time. *R.P.*

FOUR SEASONS EQUIPMENT   612-861-3838
6409 Lyndale Avenue South
Minneapolis, Minnesota 55423
Annual Sale: September. *Discount:* Some equipment is sold at less than manufacturers' suggested list price—approximately 10 percent discount. *Discount to Nonprofit Organizations:* It depends upon volume of order and the organization. *Manufacturers' Overruns and Seconds:* They have them as often as they can obtain them —10 to 25 percent discount. *Repairs:* Cross-country skis, Optimus stoves, minor sewing repairs. *Rent:* Sleeping bags, foam pads, stoves, frame packs, tents, snowshoes, cross-country skis.

HOIGAARD'S INC.   612-929-1351
3550 South Highway 100
Minneapolis, Minnesota 55416
Semiannual Sales: February and September. *Discounts:* Some merchandise is sold at less than list price. *Discount to Nonprofit Organizations:* 10 to 20 percent. *Manufacturers' Overruns and Seconds:* 40 to 50 percent discount—one of the largest discounts on this merchandise offered by any retailer in the United States;

they carry many prestigious brands, so do check. *Rent:* Snowshoes, cross-country skis. *Rental Sales:* End of season. *R.P.*

HOIGAARD'S DINKYTOWN   612-331-9360
421 14 Avenue S.E.
Minneapolis, Minnesota 55414

   Semiannual Sales: February and September. *Discount:* Some merchandise is sold at less than manufacturers' suggested list price. They have good brands here, so do check it out. *Discount to Nonprofit Organizations:* 10 to 20 percent. *Manufacturers' Overruns and Seconds:* 40 to 50 percent less than list price. *Rent:* Snowshoes, cross-country skis. *Rental Sales:* End of season. *R.P.*

MIDWEST MOUNTAINEERING   612-339-3433
309 Cedar Avenue South
Minneapolis, Minnesota 55454

   Semiannual Sales: Each spring and fall—check them for dates. *Discount:* 10 percent below manufacturers' suggested list price on brands like Fabiano, Kelty, Hunter Sleeping Bags, Camp Trails, Snow Lion, and Donner boots. *Manufacturers' Overruns and Seconds:* 30 percent below list price. *Rent:* Tents, frame packs, cross-country skis, snowshoes. *R.P.*

SKI DEN SPORTS, INC.   612-331-2310
724 Washington Avenue S.E.
Minneapolis, Minnesota 55414

   Semiannual Sales: End of winter and late summer. *Discount to Nonprofit Organizations:* Sometimes. *Repairs:* Cross-country ski equipment, downhill ski equipment, backpacks. *Rent:* Downhill and cross-country skis. *R.P.*

MOORHEAD

SUTTER'S MILL   218-233-8990
810½ Main Avenue
Moorhead, Minnesota 56560

   Discount to Nonprofit Organizations: 15 to 20 percent. *Repairs:* All sewing repairs.

ST. PAUL

EASTERN MOUNTAIN SPORTS LOB PINE   612-647-9646
1627 West Co. Road B
St. Paul, Minnesota 55113

Semiannual Sales: Late summer, late winter—check with them for this year's dates. *Discount:* Their own brand represents a saving of 10 to 20 percent off the price of comparable name brand merchandise. *Manufacturers' Overruns and Seconds:* They have them as often as they can get them. Keep your eyes open for them. *Repairs:* General repairs. *Rent:* Sleeping bags, foam pads, tents, stoves, frame packs, soft packs, snowshoes, cross-country skis. *Rental Sales:* Mid-spring and mid-fall. *R.P.*

HOIGAARD'S HIGHLAND   612-698-5521
2044 Ford Parkway
St. Paul, Minnesota 55116
   Semiannual Sales: February and September. *Discount:* Some gear is sold at less than manufacturers' suggested list price. *Discount to Nonprofit Organizations:* 10 to 20 percent. *Manufacturers' Overruns and Seconds:* 40 to 50 percent off list price; they carry many prestigious brands, so do check. *Rent:* Snowshoes, cross-country skis. *Rental Sale:* End of season—ask them for this year's date. *R.P.*

UNITED STORES   612-631-0260
Har Mar Mall
St. Paul, Minnesota 55113
   Semiannual Sales: Every spring and fall—check with them for the dates this year. They have ten stores in the Twin Cities area, so you should be able to find pretty much what you want once sales begin. *Discount:* 20 percent discount on brands like Chippewa boots, Camp Trails and Camp Ways packs, and Coleman and Hirsch Weiss (White Stag) sleeping bags. *Discount to Nonprofit Organizations:* 10 percent.

UNITED STORES   612-646-3544
1084 University Avenue
St. Paul, Minnesota 55104
   Semiannual Sales: Every spring and fall—check with them for the dates this year. *Discount:* 20 percent discount on brands like Chippewa boots, Camp Trails and Camp Ways packs, and Coleman and Hirsch Weiss (White Stag) sleeping bags. *Discount to Nonprofit Organizations:* 10 percent.

UNITED STORES   612-774-6632
1784 Montana Avenue East
St. Paul, Minnesota 55119
   Semiannual Sales: Every spring and fall—check with them for the

dates this year. *Discount:* 20 percent discount on brands like Chippewa boots, Camp Trails and Camp Ways packs, and Coleman and Hirsch Weiss (White Stag) sleeping bags. *Discount to Nonprofit Organizations:* 10 percent.

——————————————————————————— **Missouri**

CLAYTON

TAUM SAUK, INC.   314-726-0656
15 North Meramec
Clayton, Missouri 63105
  Discount to Nonprofit Organizations: 10 percent. *Rent:* Sleeping bags, tents, foam pads, stoves, frame packs, cross-country skis. *R.P.*

COLUMBIA

TAUM SAUK   314-449-1023
911 E. Broadway
Columbia, Missouri 65201
  Rent: Sleeping bags, tents, foam pads, stoves, frame packs, cross-country skis. *R.P.*

KANSAS CITY

BACK WOODS EQUIPMENT COMPANY   816-531-0200
3936 Broadway
Kansas City, Missouri 64111
  Annual Sale: February. *Discount to Nonprofit Organizations:* It depends upon the amount purchased. *Manufacturers' Seconds:* They have them occasionally, at 10 to 20 percent less than regular price. *Repairs:* Skis, stoves, packs. *Rent:* Sleeping bags, foam pads, tents, frame packs, soft packs, kayaks, cross-country skis. *Rental Sale:* February. *R.P.*

HICKMAN MILLS OUTDOOR SALES CO.   816-761-0200
11724 South 71 Highway
Kansas City, Missouri 64134
  Discount: 10 to 20 percent less than manufacturers' suggested list price on a few brands: American Camper, Coleman, Spiewak.

Discount to Nonprofit Organizations: 10 percent. *Manufacturers' Overruns:* 10 to 20 percent discount on these too.

RIDE-ON OUTDOOR SPORTS   816-753-2900
3959 Broadway
Kansas City, Missouri 64111
   Sales: Ski merchandise: mid-February; tennis goods: July; camping and backpacking equipment: October. *Discount:* They discount many items, but not all items—10 to 20 percent discount. *Discount to Nonprofit Organizations:* Depends upon group and item. *Manufacturers' Overruns and Seconds:* Up to 25 percent discount. *Clearance Center:* Up to 25 percent off on brands that may at times include Raichle, Coleman, Alpine Designs, and Gerry. *Repairs:* Bicycles, boots.

OVERLAND PARK

RIDE-ON OUTDOOR SPORTS   816-383-1911
3665 West 95 Street
Overland Park, Missouri 66206
   Sales: Ski merchandise: mid-February; tennis goods: July; camping and backpacking equipment: October. *Discount:* Many items are discounted, but not all—10 to 20 percent. *Discount to Nonprofit Organizations:* Depends upon group and item. *Manufacturers' Overruns and Seconds:* Up to 25 percent discount. *Clearance Center:* Up to 25 percent off on brands that may at times include Raichle, Coleman, Alpine Designs, and Gerry. *Repairs:* Bicycles, boots.

ST. LOUIS

OUTDOORS INC.   314-997-5866
450 N. Lindbergh Boulevard
St. Louis, Missouri 63122
   Rent: Sleeping bags, foam pads, tents, frame packs, soft packs, snowshoes, downhill and cross-country skis. *Rental Sale:* End of season. *R.P.*

WAMSER & FERMAN CO.   314-621-0480
700 North 2nd Street
St. Louis, Missouri 63102
   Manufacturers' Seconds: This store specializes in manufactur-

ers' seconds, and some used items. All are fully warranted by Wamser & Ferman. The availability of seconds depends upon the item and the manufacturer. *Rent:* Canoes, sleeping bags, tents, stoves, frame packs.

SPRINGFIELD

TAUM SAUK   417-881-3770
1453 S. Glenstone
Springfield, Missouri 65804
   Rent: Sleeping bags, foam pads, tents, stoves, frame packs, cross-country skis.

WEBSTER GROVES

MOOERS' ALPINE, LTD.   314-962-5731
14 North Gore
Webster Groves, Missouri 63119
   Annual Sale: February. *Discount to Nonprofit Organizations:* 10 percent. *Repairs:* Optimus stoves. *Rent:* Sleeping bags, foam pads, tents, frame packs, kayaks, canoes, cross-country skis. *Rental Sale:* Early fall. *R.P.*

Montana

BIGFORK

PACK & TRACK   406-837-6786
80 Electric Avenue
Bigfork, Montana 59911
   Semiannual Sales: Fall and spring—check with them for this year's dates. *Discount to Nonprofit Organizations:* 10 percent. *Clearance Center:* They have one, but the discount varies. *Repairs:* Cross-country skis, packs. *Rent:* Tents, stoves, frame packs, soft packs, snowshoes, canoes, cross-country skis. *Rental Sales:* At any time. *R.P.*

BIG SKY

MOUNTAIN DRY GOODS 406-995-4141
Big Sky, Montana 59716
 Rent: Sleeping bags, foam pads, tents, stoves, frame packs, soft packs, downhill and cross-country skis. *Rental Sales:* After each season. *R.P.*

LONE MOUNTAIN SPORTS 406-995-4471
Big Sky, Montana 59716
 Rent: Sleeping bags, foam pads, tents, stoves, frame packs, soft packs, downhill and cross-country skis. *Rental Sales:* After each season. *R.P.*

BILLINGS

MOUNTAIN CRAFT EQUIPMENT 406-259-0520
Box 30955—659 Main Street
Billings, Montana 59107
 Annual Sale: July 1. *Discount:* They sell brands like Camp 7, Galibier, and Banana Equipment at about 8 percent discount. This is one of the *very* few places in the United States where you can find a discount on Banana Equipment's excellent gear. *Discount to Nonprofit Organizations:* 20 percent. *Manufacturers' Seconds:* 20 to 30 percent discount. *Rent:* Ice axes, crampons. *Rental Sale:* April 1. *R.P.*

BOZEMAN

BEAVER POND SPORT SPECIALISTS 406-587-4261
1716 West Main Street
Bozeman, Montana 59715
 Semiannual Sales: Each spring and fall—check with them for this year's dates. *Repairs:* Minor repairs on most of what they carry. *Rent:* Sleeping bags, foam pads, tents, stoves, frame packs, soft packs, downhill and cross-country skis. *Rental Sales:* After each season. *R.P.*

HELENA

THE BASE CAMP   406-443-5360
334 North Jackson
Helena, Montana 59601
   Discount: They sell MSR, Svea, and Optimus 8-R stoves at less than manufacturers' suggested list price—one of *very* few places in the United States where you can buy MSR stoves for less than list price. *Discount to Nonprofit Organizations:* 10 percent. *Manufacturers' Seconds:* 10 percent discount. *Repairs:* Sewing repairs, cross-country skis. *Rent:* Sleeping bags, tents, stoves, frame packs, soft packs, snowshoes, ice axes, crampons. *Rental Sales:* End of each season—ask them for this year's dates. *R.P.*

THE SPORTS CHALET   406-442-2790
21 North Last Chanch Gulch
Helena, Montana 59601
   Annual Sale: April. *Discount to Nonprofit Organizations:* 5 to 20 percent. *Manufacturers' Overruns and Seconds:* When they can get them they sell for 30 to 40 percent below list price. *Repairs:* Bicycles, skateboards, packs, stoves, tents, boats. *Rent:* Downhill and cross-country skis. *Rental Sale:* End of winter season—ask them for this year's date. *R.P.*

KALISPELL

ROCKY MOUNTAIN OUTFITTER   406-755-2442
135 Main Street
Kalispell, Montana 59901
   Discount to Nonprofit Organizations: 10 to 20 percent. *Repairs:* Cross-country skis, plus sewing repairs and other minor gear repairs. *Rent:* Tents, stoves, frame packs, soft packs, snowshoes, ice axes, crampons, cross-country skis. *Rental Sales:* Seasonally. *R.P.*

REEDS SPORTING GOODS   406-755-4617
132 Main
Kalispell, Montana 59901
   Discount: 10 percent off manufacturers' suggested list on brands like Camp Trails, Coleman, White Stag. *Discount to Non-*

profit Organizations: 20 percent. *Repairs:* Camp stoves, rods and reels.

MISSOULA

THE TRAIL HEAD   406-543-6966
501 South Higgins Avenue
Missoula, Montana 59801
   Manufacturers' Overruns and Seconds: 10 to 20 percent discount. *Clearance Center:* 25 percent discount on brands that may include Jansport, Trailwise, Camp 7, and Snow Lion. *Repairs:* Stoves, skis, packs. *Rent:* Sleeping bags, foam pads, tents, stoves, frame packs, soft packs, snowshoes, ice axes, crampons, cross-country skis. *R.P.*

WEST YELLOWSTONE

BUD LILLY'S TROUT SHOP   406-646-7801
Box 698
West Yellowstone, Montana 59758
   Special Features: Bud Lilly's has the largest collection of hand-tied flies in the state. They're not a discounter, but this is an interesting shop.

————————————————————————— **Nebraska**

BEATRICE

TAYLOR RENTAL CENTER   402-223-5261
1903 North 6th Street
Beatrice, Nebraska 68310
   Discount: 20 percent off list price on brands like Coleman and Wenzl tents. *Manufacturers' Overruns and Seconds:* 20 percent discount. *Rent:* Sleeping bags, tents, stoves, bicycles, ice chests, Thermos jugs, lanterns. *R.P.:* One-half is applicable.

LINCOLN

BIVOUAC   308-432-0090
1235 Q Street
Lincoln, Nebraska 68508
  Discount: Up to 10 percent discount on most gear; they handle
brands like Fabiano, Trailwise, Jansport, and Snow Lion. *Rent:*
Sleeping bags, tents, stoves, frame packs, cross-country skis. *Rental
Sales:* After Christmas, after summer.

OMAHA

BACK WOODS EQUIPMENT CO.   402-345-0303
3724 Farnam
Omaha, Nebraska 68131
  Rent: Sleeping bags, pads, tents, frame packs, soft packs, kayaks,
cross-country skis. *Rental Sale:* February. *R.P.*

SIDNEY

CABELA'S, INC.   308-254-5505
812–13 Avenue
Sidney, Nebraska 69162
  Clearance Center: 50 percent off on brands that may include
Camp Trails, Raichle, Vasque, Woolrich.

——————————————————————— **New Hampshire**

CLAREMONT

CLEAR MOUNTAIN SPORTS   603-542-4341
44 Tremont Square
Claremont, New Hampshire 03743
  Discount: 10 percent less than manufacturers' suggested list
price on brands like Alpine Designs, Jansport, Sierra Designs. *Discount to Nonprofit Organizations:* 10 percent. *Rent:* Cross-country
skis. *R.P.*

DUBLIN

SUMMERS SKI & MOUNTAIN CENTER   603-563-8556
Route 101
Dublin, New Hampshire 03444
Discount: 10 percent discount on most name brand merchandise. *Rent:* Cross-country skis, tents, frame packs, stoves, sleeping bags, foam pads, canoes, kayaks, crampons, ice axes, snowshoes.

DURHAM

WILDERNESS TRAILS   603-868-5584
12 Pettee Brook Lane
Durham, New Hampshire 03824
Discount to Nonprofit Organizations: 10 to 20 percent. *Repairs:* Optimus stoves, cross-country skis. *Rent:* Sleeping bags, foam pads, tents, stoves, frame packs, soft packs, snowshoes, cross-country skis. *Rental Sale:* March. *R.P.*

HANOVER

DARTMOUTH CO-OP   603-643-3100
Hanover, New Hampshire 03755
Discount: 10 percent discount on all merchandise to all co-op members; membership is $10 per year and entitles one to a special members' sale day.

HENNIKER

POLE & PEDAL   603-428-3242
Bridge Street—P.O. Box 327
Henniker, New Hampshire 03242
Annual Sale: Each spring—check with them for this year's date. *Discount to Nonprofit Organizations:* 20 percent. *Rent:* Foam pads, tents, frame packs, soft packs, snowshoes, kayaks, canoes, bicycles, ice axes, cross-country skis. *Rental Sale:* End of each season —ask for this year's dates. *R.P.*

LITTLETON

STOD NICHOLS, INC.   603-444-5597
43–45 Main Street
Littleton, New Hampshire 03561
  Semiannual Sales: First weekend in August, first weekend in
January. *Discount to Nonprofit Organizations:* 10 percent, but only
to schools.

MANCHESTER

BI-RITE MERCHANDISERS   603-669-1340
South State
Manchester, New Hampshire 03101
  Discount: At this mail-order-merchandise discount center, prices
are about 20 percent less than list price on all general sporting
goods. It's worth writing for the catalog.

NORTH CONWAY

EASTERN MOUNTAIN SPORTS, INC.   603-356-5433
Main Street
North Conway, New Hampshire 03860
  Semiannual Sales: April and October, with up to 40 percent
savings. *Discount:* Their own brand of merchandise costs about 10
to 20 percent less than comparable brand name merchandise. *Discount to Nonprofit Organizations:* Varies according to group. *Manufacturers' Overruns and Seconds:* 20 to 40 percent less than list
price. *Rent:* Foam pads, tents, climbing boots, frame packs, soft
packs, cross-country skis, ice axes, crampons. *R.P.*

GRALYN SPORTS CENTER   603-356-5546
North Conway, New Hampshire 03860
  Manufacturers' Seconds: They carry them but didn't tell us what
their discount schedule is. *Rent:* Kayaks, canoes, bicycles. *Rental*

Sale: At any time. *R.P. Special Features:* These people are canoe specialists.

INTERNATIONAL MOUNTAIN EQUIPMENT   603-356-5287
Box 494—Main Street
North Conway, New Hampshire 03860

Semiannual Sales: February and September. *Discount to Non-profit Organizations:* Up to 20 percent discount. *Manufacturers' Overruns and Seconds:* Up to 50 percent off list price, when they can get them. *Repairs:* Torn tents, parkas, pants, etc. *Rent:* Frame packs, tents, snowshoes, ice axes, cross-country skis. *Rental Sale:* Each spring—ask them for this year's dates. *R.P. Special Features:* They take custom orders for Gore-Tex® jackets and bivvy bags.

NORTH WOODSTOCK

SKIMEISTER SPORT SHOP   603-745-2767
Main Street
North Woodstock, New Hampshire 03262

Semiannual Sales: Spring and fall—check for exact dates. Skimeister is particularly good if you're looking for something like double boots on sale. *Repairs:* Minor repairs. *Rent:* Tents, stoves, frame packs, snowshoes, ice axes, crampons, cross-country skis. *R.P.*

PETERBOROUGH

EASTERN MOUNTAIN SPORTS   603-924-7276
Vose Farm Road
Peterborough, New Hampshire 03458

Semiannual Sales: October and end of April. *Discount:* Their own brand of gear generally represents a savings of 20 percent over comparable name brand merchandise. *Manufacturers' Overruns and Seconds:* 15 to 20 percent less than list price. *Rent:* Cross-country skis. *Rental Sale:* October and end of April.

BASKING RIDGE

OVERALL OUTFITTER   201-766-6521
24 West Oak Street
Basking Ridge, New Jersey 08812
   Discount: Merchandise is sold about 10 percent below suggested list price. *Manufacturers' Overruns and Seconds:* 20 percent below list price. *Clearance Center:* 30 percent below list price. *Repairs:* Packs and stoves. *Rent:* Tents, stoves, frame packs, soft packs, snowshoes, ice axes, crampons, cross-country skis. *Rental Sales:* Negotiable. *R.P.*

CLARK

HILLS & TRAILS   201-574-1240
93 Brant Avenue (Garden State Parkway Interchange)
Clark, New Jersey 07066
   Semiannual Sales: Fall and spring—check for this year's dates. *Discount to Nonprofit Organizations:* 5 to 20 percent. *Manufacturers' Overruns and Seconds:* They occasionally have them, 15 to 25 percent below list price. *Repairs:* Skis, boots, some sewing. *Rent:* Sleeping bags, foam pads, tents, stoves, frame packs, soft packs, snowshoes, downhill and cross-country skis, ice axes, and crampons. *Rental Sales:* Seasonally—ask them for this year's dates. *R.P.*

EAST BRUNSWICK

DOMINO'S SPORT CENTER   201-257-4807
421 Ryders Lane
East Brunswick, New Jersey 28826
   Discount: 10 to 25 percent off list price on a few brands like Hunter Sleeping Bags, Mountain Products backpacks, and Coleman and Comfy parkas. *Clearance Center:* Up to 50 percent off list price.

GREEN BROOK

OVERALL OUTFITTER   201-968-4230
62 Route 22
Green Brook, New Jersey 08812
   Semiannual Sales: Spring and fall—check with them for dates.
Discount to Nonprofit Organizations: 10 percent. *Manufacturers'
Overruns and Seconds:* 15 to 20 percent below list price. *Repairs:* Stoves, light sewing repairs.

LAWRENCEVILLE

HERMAN'S WORLD OF SPORTING GOODS   609-799-3000
Quaker Bridge Mall
Lawrenceville, New Jersey 08648
   Annual Sale: November 28 or thereabouts *Discount:* Generally,
they pass on a 5 to 10 percent savings on items like Coleman fuel
and Coleman freezers.

LEDGEWOOD

RAMSEY OUTDOOR STORE, INC.   201-584-7799
Route 46
Ledgewood, New Jersey 07852
   Discount: 10 to 15 percent below list price on brands like Dunham, Chippewa boots, Dexter boots, Raichle, Trailwise, Coleman,
White Stag, Camp Trails, and Mountain Products.

LONG VALLEY

WILDERNESS SHOP   201-876-4648
9 West Mill Road (Route 513)
Long Valley, New Jersey 07853
   Annual Sale: February or March. *Discount to Nonprofit Organizations:* 10 to 20 percent. *Manufacturers' Seconds:* 20 percent off
list price. *Repairs:* Any type of sewing repair. *Rent:* Sleeping bags,
foam pads, tents, stoves, frame packs, soft packs, snowshoes, ice
axes, crampons, cross-country skis. *R.P.*

PARAMUS

HERMAN'S WORLD OF SPORTING GOODS   201-843-1000
Garden State Shopping Center
Routes 4 and 17
Paramus, New Jersey 07652
   Annual Sale: November 28 or thereabouts. *Discount:* Generally, they pass on a 5 to 10 percent discount on items like Coleman fuel and Coleman freezers.

RAMSEY OUTDOOR STORE, INC.   201-261-5000
226 State Highway 17
Paramus, New Jersey 07652
   Discount: 10 to 15 percent below list price on brands like Dunham, Chippewa boots, Dexter boots, Raichle, Trailwise, Coleman, White Stag, Camp Trails, and Mountain Products.

PRINCETON

THE NICKEL   609-924-3001
354 Nassau Street
Princeton, New Jersey 08540
   Annual Sale: Beginning of February. *Discount to Nonprofit Organizations:* Varies—check with them. *Manufacturers' Overruns and Seconds:* 30 to 40 percent below list price. *Repairs:* Minor repairs on merchandise bought here. *Rent:* Sleeping bags, foam pads, parkas, tents, stoves, frame packs, soft packs, snowshoes, cross-country skis. *Rental Sale:* At any time. *R.P.*

RAMSEY

RAMSEY OUTDOOR STORE, INC.   201-327-8141
835 Route 17
Ramsey, New Jersey 07446
   Discount: 10 to 15 percent below list price on brands like Dunham, Chippewa boots, Dexter boots, Raichle, Trailwise, Coleman, White Stag, Camp Trails, and Mountain Chalet.

ALBUQUERQUE

MOUNTAINS & RIVERS   505-268-4876
2320 Central S.E.
Albuquerque, New Mexico 87106
   Semiannual Sales: March or April, and September. *Discount to Nonprofit Organizations:* 10 percent. *Repairs:* Boots, jackets, tents. *Rent:* Sleeping bags, foam pads, tents, frame packs, snowshoes, canoes, cross-country skis. *R.P.*

MOUNTAIN CHALET   505-881-5223
6307 Menaul Boulevard N.E.
Albuquerque, New Mexico 87110
   Rent: Sleeping bags, foam pads, tents, frame packs, soft packs, kayaks, cross-country skis. *Rental Sale:* February. *R.P.*

SANDIA MOUNTAIN OUTFITTERS   505-293-9725
9611 Menaul Boulevard N.E.
Albuquerque, New Mexico 87112
   Rent: Sleeping bags, foam pads, tents, frame packs, soft packs, snowshoes, ice axes, crampons, cross-country skis.

THE WILDERNESS CENTRE   505-266-8113
2421 San Pedro Drive N.E.
Albuquerque, New Mexico 87110
   Discount to Nonprofit Organizations: 10 percent. *Repairs:* Tents, stoves, packs, torn nylon. *Rent:* Sleeping bags, tents, frame packs, snowshoes, cross-country skis. *R.P.*

SANTA FE

H. COOK SPORTING GOODS   505-988-4466
De Vargas Mall
Santa Fe, New Mexico 87501
   Clearance Center: They do have one, with brands that will at times include Raichle, Alpine Designs, Coleman, and Gerry, but didn't tell us what reductions they generally give. *Repairs:* Skis.

Rent: Downhill and cross-country skis. *Rental Sale:* End of ski season. *R.P.*

PEAK & PLAIN OUTFITTERS   505-982-8948
426 Abeyta Street—Box 2538
Santa Fe, New Mexico 87501

Discount to Nonprofit Organizations: It varies. *Clearance Center:* They have one, with brands that will at times include Camp Trails, Ascente, Petzoldt, and their own High Camp Designs, but they didn't tell us what the reductions generally are—worth checking.

─────────────────────────────── **New York**

ALBANY

ADIRONDACK DAN ARMY-NAVY   518-434-3495
Steuben & James
Albany, New York 12207

Discount: 25 to 60 percent below manufacturers' list price on brands that include Dunham boots, American Footwear, Hunter Sleeping Bags, Greylock Mountain Industries (sleeping bags), Woods products, and their own Academy brand. *Special Features:* They carry nylon cloth for the home sewer. They have a good attitude: "Since 1904 our attitude has been low initial investment, and anyone can afford to camp".

ARDSLEY

EASTERN MOUNTAIN SPORTS   914-693-6160
725 Saw Mill River Road
Ardsley, New York 10502

Semiannual Sales: End of October and April or May. *Discount:* Their own EMS brands represent a saving of from 10 to 20 percent compared with similar brand name merchandise. *Clearance Center:* They have a bargain basement here that is one of the consistently best sources of bargain outdoor gear we've ever seen—up to 70 percent off on brands like Snow Lion, North Face, Molitor, Kelty, and Optimus. There's a separate entrance to the bargain basement, outside and to the left of the store. If you can't find it, ask. *Re-*

pairs: Minor. *Rent:* Sleeping bags, foam pads, Mickey Mouse boots, tents, stoves, frame packs, soft packs, snowshoes, downhill skis, ice axes, crampons, cross-country skis, ski mountaineering skis. *Rental Sales:* Twice each year, at the end of summer and winter seasons. *R.P.*

ARMONK

KREEGER & SON   914-273-8520
387 Main Street
Armonk, New York 10504
   Discount to Nonprofit Organizations: They'll give you a break only if it's a large bulk order. Ask.

BINGHAMTON

EUREKA CAMPING CENTER   607-723-4179
625 Conklin Road
Binghamton, New York 13902
   Discount: 10 percent below manufacturers' suggested retail price on brands like Camp Trails, Class 5, Gerry, Donner, Vasque, North Face, and Woolrich. *Manufacturers' Overruns and Seconds:* 10 to 25 percent less than list price *Clearance Center:* They have one sometimes. Check when you are in the area. *Repairs:* Tents. *Rent:* Tents, frame packs, canoes, cross-country skis. *R.P.*

NIPPENOSE EQUIPMENT   607-724-4363
Stephens Square—81 State Street
Binghamton, New York 13901
   Discount to Nonprofit Organizations: This must be arranged through their main office in Williamsport, Pennsylvania (225 West 4th Street, Williamsport, Pennsylvania 17701). *Manufacturers' Seconds:* 20 to 40 percent off list price. *Repairs:* Basic sewing and grommet repairs.

BROCKPORT

TRACK & TRAIL SKI SHOP INC.   716-637-2910
82 Main
Brockport, New York 14420
   Rent: Tents, downhill skis, cross-country skis.

BUFFALO

BROWNIE'S ARMY & NAVY STORE, INC.   716-854-2218
575 Main Street
Buffalo, New York 14203
   Discount: Up to 10 percent discount on brands like Woolrich, White Stag, American Footwear, Dunham. *Manufacturers' Overruns and Seconds:* 10 to 30 percent less than list price.

HERMAN'S WORLD OF SPORTING GOODS   716-853-8870
699 Main Street
Buffalo, New York 14203
   Annual Sale: November 28 or thereabouts—watch for unscheduled sales. *Discount:* Generally they pass on a 5 to 10 percent discount on items like Coleman fuel and Coleman freezers.

CARLE PLACE

EASTERN MOUNTAIN SPORTS   516-747-7360
174 Glen Cove Road
Carle Place, Long Island, New York 11514
   Semiannual Sales: End of October and late April or the beginning of May. *Discount:* Their own EMS brand represents a saving of from 10 to 20 percent off the cost of comparable brand name merchandise. *Repairs:* Minor repairs. *Rent:* Sleeping bags, foam pads, Mickey Mouse boots, tents, stoves, frame packs, soft packs, snowshoes, cross-country skis. *Rental Sales:* End of October, end of April—ask them. *R.P.*

FAYETTEVILLE

HERMAN'S WORLD OF SPORTING GOODS   315-637-5081
Fayetteville Mall
Route 5—East Genesee Street
Fayetteville, New York 13066
   Annual Sale: November 28 or thereabouts—watch for unscheduled sales. *Discount:* Generally they pass on a 5 to 10 percent discount on items like Coleman fuel and Coleman lanterns.

GLENS FALLS

INSIDE EDGE   518-793-5676
624 Glen Street
Glens Falls, New York 12801
   Discount to Nonprofit Organizations: It varies; check with them.
Clearance Center: 50 percent off on brands that may include Class
5, North Face, Alpine Designs, Sorel boots, and Nordica; they are
very much worth checking as often as you're in the area. *Repairs:* Skis and bicycles.

RELIABLE RACING SUPPLY   518-793-5677
624 Glen Street
Glens Falls, New York 12801
   Clearance Center: Up to 50 percent off list price. *Discount to
Nonprofit Organizations:* Negotiable. *Rent:* Tents, snowshoes,
downhill and cross-country skis, bicycles. *Rental Sale:* Each summer. Ask them for this year's date. *R.P.*

GREECE

HERMAN'S WORLD OF SPORTING GOODS   716-227-2010
Greece Town Hall
2213 Ridge Road West
Greece, New York 14626
   Annual Sale: Around November 28—watch for unscheduled
sales. *Discount:* They sell Coleman products at about 5 to 10 percent discount.

HUNTINGTON STATION

HERMAN'S WORLD OF SPORTING GOODS   516-423-1700
Route 110
Korvette Shopping Center
Huntington Station, New York 11746
   Annual Sale: Around November 28—watch for unscheduled
sales. *Discount:* They sell Coleman products at about 5 to 10 percent discount.

ITHACA

NIPPENOSE EQUIPMENT   607-272-6868
Dewitt Mall—215 North Cayuga
Ithaca, New York 14850

Discount to Nonprofit Organizations: Must be arranged with
their main office: 225 West Fourth Street, Williamsport, Pennsylvania 17701. *Manufacturers' Seconds:* 20 to 40 percent off list
price. *Repairs:* Basic sewing and grommet repairs.

LAKE PLACID

EASTERN MOUNTAIN SPORTS   518-523-2505
Main Street
Lake Placid, New York 12946

Sales: Mid-January, April, July, and August (two weeks), and
mid-October. *Discount:* They give bulk discounts of 5 percent on
orders over $250, 10 percent on orders over $500. If you're buying more than that, see Dave Cilley and maybe you'll be able to
arrange a larger discount. Also, EMS's own brand of gear usually
represents a saving of between 10 and 20 percent over comparable merchandise. *Discount to Nonprofit Organizations:* Same as
regular bulk discount outlined above. *Manufacturers' Overruns and
Seconds:* 30 to 70 percent off list price. *Repairs:* Stoves only, no
sewing. *Rent:* Sleeping bags, foam pads, parkas, Mickey Mouse
boots, stoves, frame packs, soft packs, snowshoes, ice axes, cross-
country skis, baby carriers, lanterns. *Rental Sale:* Spring and fall—
check with them for this year's dates. *R.P.*

LATHAM

HANSON'S TRAIL NORTH   518-785-0340
895 New Loudon Road
Latham, New York 12110

Annual Sale: March. *Discount to Nonprofit Organizations:* 20
percent. *Repairs:* Skis. *Rent:* Sleeping bags, tents, frame packs,
snowshoes, ice axes, crampons, cross-country skis. *Rental Sale:*
March.

LIVERPOOL

LIVERPOOL SPORT CENTER　　　　　　315-457-2290
125 First Street
Liverpool, New York 13088
　Semiannual Sales: February and August. *Discount to Nonprofit Organizations:* 15 percent. *Manufacturers' Overruns:* 30 to 50 percent off list price. *Repairs:* Minor repairs. *Rent:* Tents, stoves, frame packs, soft packs, snowshoes, cross-country skis, downhill skis, kayaks, canoes, bicycles, crampons. *Rental Sale:* End of season—ask them for this year's dates. *R.P.*

MILLERTON

TACONIC SPORTS & CAMPING CENTER　　518-789-3288
R.D. #2, Rudd Pond Road
Millerton, New York 12546
　Discount: They do sell at less than list price but didn't tell us how much less. They are worth checking, as they carry a few excellent brands, such as Alpine Designs and Woolrich. *Discount to Nonprofit Organizations:* It varies. *Repairs:* Minor repairs.

MINEOLA

E-Z RENTS EVERYTHING　　　　　　516-248-1230
155 Jericho Turnpike
Mineola, New York 11501
　Annual Sale: Early spring. *Discount:* They have a few brands like Camel Tents and Trailblazer backpacks that they sell at least 50 percent below manufacturers' price. *Rent:* Sleeping bags, tents, stoves, frame packs, shovels, heaters, canopies, lanterns, ski carriers, bicycle carriers, audio-visual accessories, exercise machines.

NEW YORK CITY

HERMAN'S WORLD OF SPORTING GOODS　　212-688-4603
845 Third Avenue
New York, New York 10022
　Annual Sale: Around November 28—watch for other un-

scheduled sales. *Discount:* They sell Coleman products at about 5 to 10 percent discount.

HERMAN'S WORLD OF SPORTING GOODS   212-730-7400
135 West 42nd Street
New York, New York 10036

Annual Sale: Around November 28—watch for other unscheduled sales. *Discount:* They sell Coleman products at about 5 to 10 percent discount.

KREEGER & SONS   212-575-7825
16 West 46th Street
New York, New York 10036

Discount: If you are buying over $250 worth of merchandise ask about a discount for bulk orders. *Discount to Nonprofit Organizations:* Only on bulk orders. *Clearance Center:* 20 percent discount on occasion. *Rent:* Cross-country skis. *Repairs:* Everything except canvas tents.

PARAGON SPORTING GOODS   212-255-8036
871 Broadway (at 18th Street)
New York, New York 10003

Sales: February, May, June, July, September. *Discount:* At least 10 percent on brands like Alpina, Iowa, Trailwise, Class 5, High and Light, Coleman, Kelty, Gerry, Wilderness Experience, Hine/Snowbridge, and Fabiano, making this one of the best discounters in the New York City area. *Manufacturers' Overruns and Seconds:* 20 to 30 percent off list price.

SCANDINAVIAN SKI SHOP   212-757-8524
40 West 57th Street
New York City, New York 10020

Rent: Cross-country and downhill skis.

SPORTIVA SPORTHAUS   212-734-7677
1653 2nd Avenue (at 86th Street)
New York, New York 10028

Rent: Cross-country and downhill skis.

STANLEY'S ARMY & NAVY STORES, INC.   212-158-2846
90-65 Sutphin Boulevard
Jamaica, New York 11435

Discount to Nonprofit Organizations: 10 percent. *Manufacturers' Seconds:* Between 30 and 40 percent off list price on brands like Timberland, Benjack Packs, Coleman, and Spiewak parkas. *Rent:* Sleeping bags, tents, stoves, air mattresses, folding cots, lanterns.

PLAINVIEW

A TO Z RENTAL CENTER   516-293-4192
1526 Old Country Road
Plainview, New York 11803
    Discount to Nonprofit Organizations: 10 percent discount on a few less than fancy but serviceable brands: Trailblazer, Camel, Canyon. *Rent:* Sleeping bags, tents, stoves, frame packs. *Rental Sale:* Fall. *R.P.*

ROCHESTER

ABC SPORT SHOP   716-271-4550
185 Norris Drive
Rochester, New York 14610
    Annual Sale: October. *Discount to Nonprofit Organizations:* 10 to 20 percent. *Repairs:* Hiking boots, fishing reels, tents. *Rent:* Foam pads, tents, stoves, frame packs, soft packs, snowshoes, canoes, ice axes, crampons, cross-country skis. *Rental Sale:* October. *R.P. Special Features:* They carry a very good selection of tents.

SNOW COUNTRY   716-586-6460
3330 Monroe Avenue
Rochester, New York 14618
    Semiannual Sales: Late August, late February. *Discount to Nonprofit Organizations:* 20 percent to schools only. *Clearance Center:* 20 to 60 percent off list price on brands that may at times include Vasque, Gerry, Class 5, and North Face. *Repairs:* Alpine and Nordic ski equipment. *Rent:* Cross-country skis. *Rental Sale:* Each spring—ask for the date. *R.P.*

SYRACUSE

NIPPENOSE EQUIPMENT COMPANY   315-446-3838
3006 Erie Boulevard East
Syracuse, New York 13224
    Discount to Nonprofit Organizations: Contact their main office in

Pennsylvania first: 225 West Fourth Street, Williamsport, Pennsylvania 17701.

TARRYTOWN

BACKCOUNTRY OUTFITTERS   914-631-0409
625 White Plains Road
Tarrytown, New York 10591
   Discount to Nonprofit Organizations: It depends upon volume of sale. *Manufacturers' Seconds:* Sometimes, 20 percent less than list price. *Clearance Center:* 20 percent off on brands that may at times include Jansport, Class 5, and Mountain-10.

TONAWANDA

EASTERN MOUNTAIN SPORTS   716-838-4200
1270 Niagara Falls Boulevard
Tonawanda, New York 14150
   Semiannual Sales: November and May. *Discount:* Their own brand of merchandise costs about 10 to 20 percent less than comparable name brand merchandise. *Discount to Nonprofit Organizations:* Large orders only. *Manufacturers' Overruns and Seconds:* 10 to 30 percent off list price. *Rent:* Cross-country skis. R.P.

WEBSTER

TAYLOR RENTAL CENTER   716-872-2770
150 Orchard Street
Webster, New York 14580
   Rent: Sleeping bags, foam pads, tents, stoves, frame packs, canoes, lanterns, cots, heaters, portable refrigerators.

YONKERS

HERMAN'S WORLD OF SPORTING GOODS   914-423-5400
Cross-County Shopping Center
Major Deegan Expressway & Cross-County Parkway
Yonkers, New York 10704

Annual Sale: Around November 28—watch for other un-scheduled sales. *Discount:* They sell Coleman products at about 5 to 10 percent discount.

--- **North Carolina**

BOONE

FOOTSLOGGERS   704-264-6565
204 Blowing Rock Road
Boone, North Carolina 28607
   Discount: They sometimes sell gear at up to 20 percent below list price. *Discount to Nonprofit Organizations:* 10 to 25 percent. *Manufacturers' Overruns:* They have them occasionally, at 20 percent below list price. *Clearance Center:* 20 to 40 percent off list price on some very good brands. *Rent:* Sleeping bags, tents, frame packs, canoes. *Rental Sale:* October. *R.P.*

BRYSON CITY

FOLKESTONE LODGE   704-488-2730
Route 1—Box 310—West Deep
Creek Road
Bryson City, North Carolina 28713
   Rent: Sleeping bags, foam pads, tents, frame packs, bicycles, soft packs.

CHARLOTTE

ALANBY, INC.   704-568-8048
3040-B Eastway Drive
Charlotte, North Carolina 28205
   Semiannual Sales: Right after Christmas, and July 4. *Repairs:* Backpacks. *Rent:* Backpacks, sleeping bags, tents, cross-country skis, snowshoes, canoes. *Special Features:* They will custom design any soft goods. Also, they sell Gore-Tex® by the yard.

JESSE BROWN'S BACKPACKING,
MOUNTAINEERING, CANOEING,
KAYAKING   704-568-2152
2843 Eastway Drive
Charlotte, North Carolina 28205

Sales: Between Christmas and New Year, also spring and fall. *Discount to Nonprofit Organizations:* 5 to 20 percent. *Clearance Center:* A very small one, offering 10 to 50 percent reductions. *Repairs:* Minor repairs on stoves, packframes, etc. *Rent:* Sleeping bags, foam pads, tents, stoves, frame packs, canoes. *Rental Sale:* Canoes are sold at the end of the summer season. *R.P.*

CULLOWHEE

CULLOWHEE WILDERNESS OUTFITTER   704-293-9741
P.O. Box V
Alpine Building—Highway 107
Cullowhee, North Carolina 28723

Discount to Nonprofit Organizations: 10 percent. *Manufacturers' Seconds:* They have seconds of canoes at 15 to 20 percent less than list price. *Rent:* Sleeping bags, foam pads, parkas, tents, stoves, frame packs, soft packs, kayaks, canoes. *Rental Sales:* November and March. *R.P.*

DURHAM

APPALACHIAN OUTFITTERS   919-489-1207
2805 Hope Valley Road
Durham, North Carolina 27707

Discount: 5 to 15 percent below manufacturers' suggested list prices on brands like North Face, Sierra Designs, Alpenlite, Camp 7, and Gerry. *Repairs:* They repair stoves. *Rent:* Sleeping bags, foam pads, stoves, frame packs, hiking boots, cross-country skis, canoes. *Rental Sale:* First week of December. *R.P.*

GREENSBORO

BLUE RIDGE MOUNTAIN SPORTS   918-275-8115
1507 Spring Garden Street
Greensboro, North Carolina 27403

Discount to Nonprofit Organizations: It varies; speak with Jim Neese about it. *Rent:* Sleeping bags, foam pads, tents, stoves, frame packs, soft packs, kayaks, canoes. *Rental Sale:* January. *R.P.:* They'll apply half of the rental fee toward the purchase price.

CAROLINA OUTDOOR SPORTS   919-274-1862
844 West Lee Street
Greensboro, North Carolina 27403
   Discount to Nonprofit Organizations: It varies, but Greensboro is no place to get much of a bargain on anything. *Manufacturers' Seconds:* 10 to 15 percent discount, when they get them. *Repairs:* All types of sewing repairs. *Rent:* Sleeping bags, foam pads, tents, stoves, frame packs, kayaks, canoes, ice axes, crampons. *R.P.*

RALEIGH

CAROLINA OUTDOOR SPORTS   919-782-8288
1520 Dixie Trail
Raleigh, North Carolina 27607
   Annual Sale: Right after Christmas. *Discount to Nonprofit Organizations:* 5 to 10 percent, depending upon the group. *Repairs:* Tents, packs, sleeping bags. *Rent:* Sleeping bags, foam pads, tents, stoves, frame packs, canoes. *Rental Sale:* After Christmas. *R.P.*

WINSTON-SALEM

APPALACHIAN OUTFITTERS   919-784-7402
4240 Kernersville Road
Winston-Salem, North Carolina 27107
   Discount: 5 to 15 percent on some very good brands, including North Face, Gerry, Camp 7, and Jansport. *Repairs:* Stove repairs. *Rent:* Sleeping bags, foam pads, stoves, frame packs, hiking boots, cross-country skis. *Rental Sale:* First week of December. *R.P.*

TATUM OUTFITTERS   919-748-1718
1215 Link Road
Winston-Salem, North Carolina 27103
   Annual Sale: Right after Christmas. *Discount to Nonprofit Or-*

ganizations: 10 percent on quantity purchases. *Repairs:* Minor repairs on all equipment.

North Dakota

FARGO

MOUNTAIN SPECIALTIES   701-235-4007
19 South 8th Street
Fargo, North Dakota 58102
   Discount: Some merchandise is discounted, but only when the suggested markup exceeds 40 percent. *Discount to Nonprofit Organizations:* 10 to 30 percent, depending on the amount of the purchase. *Clearance Center:* 20 to 80 percent on top brands. *Repairs:* Cross-country skis. *Rent:* Tents, stoves, snowshoes, kayaks, cross-country skis. *Rental Sale:* Skis in late winter.

GRAND FORKS

GALLEON GRAND FORKS   701-772-7893
2010 13th Avenue North
Grand Forks, North Dakota 58201
   Annual Sale: September. *Discount:* 15 to 25 percent on brands like Mountain-10 boots, Himalayan Industries backpacks, and Himalayan Industries and Pine Line sleeping bags. *Discount to Nonprofit Organizations:* 25 to 40 percent off list price—one of the largest discounts in the United States.

Ohio

AKRON

CAMP & TRAIL HUTTE   216-535-1225
469 East Exchange
Akron, Ohio 44304

Semiannual Sales: July and February. *Discount:* 10 to 12 percent below manufacturers' suggested list price on equipment from Alpine Products, Summit Boots, Galibier, and Mountain Products; this is also one of the few places in the United States where you can get any discount on Lowe Alpine Systems' excellent gear. *Rent:* Sleeping bags, parkas, tents, stoves, frame packs, soft packs, ice axes, snowshoes, crampons, cross-country skis. *Rental Sales:* July and February. *R.P.*

MOUNTAIN SPORTS   216-867-7634
1698 Merriman Road
Akron, Ohio 44313
   Annual Sale: February. *Discount to Nonprofit Organizations:* 10 to 20 percent. *Repairs:* Downhill and cross-country skis, bicycles. *Rent:* Sleeping bags, foam pads, tents, stoves, frame packs, cross-country skis.

CINCINNATI

APPALACHIAN OUTFITTERS   513-752-3032
951 Batavia Pike
Cincinnati, Ohio 45245
   Discount: 5 to 15 percent on North Face, Gerry, Jansport, Camp 7, and other quality gear. *Repairs:* Stoves. *Rent:* Sleeping bags, foam pads, stoves, frame packs, hiking boots, cross-country skis. *Rental Sale:* First week in December. *R.P.*

GOVERNMENT SURPLUS DEPOT, INC.   513-541-8700
4031 Hamilton Avenue
Cincinnati, Ohio 45223
   Discount to Nonprofit Organizations: It depends upon organization and size of the order. They carry military discount gear from several nations.

OUTDOOR ADVENTURES   513-281-2594
39 Calhoun Street
Cincinnati, Ohio 45219
   Semiannual Sales: They have end-of-season sales. Check with them for this year's dates. *Discount to Nonprofit Organizations:* The amount depends upon the quantity of the order. *Repairs:* Small equipment repairs; if they can't fix it, they'll tell you where you can have it done. *Rent:* Foam pads, tents, stoves, frame packs, soft

packs, kayaks. *Rental Sales:* End of season—check with them for this year's dates. *R.P.*

WILDERNESS OUTFITTERS, INC.   513-931-1470
7619 Hamilton Avenue
Cincinnati, Ohio 45231
   Discount: They do sell at less than manufacturers' suggested list price but didn't specify how much less. *Manufacturers' Overruns and Seconds:* 10 to 20 percent off list price. *Rent:* Cross-country skis.

CLEVELAND

ADLER'S FAMOUS OUTFITTERS   216-696-5222
728 Prospect S.E.
Cleveland, Ohio 44115
   Semiannual Sales: April and August. *Discount:* 10 percent below list price on some items. *Discount to Nonprofit Organizations:* 5 to 10 percent. *Manufacturers' Overruns and Seconds:* 20 to 50 percent off list price. *Repairs:* Stoves, lanterns, cross-country skis. *Rent:* Sleeping bags, foam pads, tents, stoves, frame packs, snowshoes, ice axes, cross-country skis. *Rental Sale:* Each spring— ask them for the date this year. *R.P.*

ADLER'S FAMOUS OUTFITTERS   216-382-7282
4504 Mayfield
Cleveland, Ohio 44121
   Semiannual Sales: April and August. *Discount:* 10 percent below list price on some items. *Discount to Nonprofit Organizations:* 5 to 10 percent. *Manufacturers' Overruns and Seconds:* 20 to 50 percent off list price. *Repairs:* Stoves, lanterns, cross-country skis. *Rent:* Sleeping bags, foam pads, tents, stoves, frame packs, snowshoes, ice axes, cross-country skis. *Rental Sale:* Each spring— ask them for this year's date. *R.P.*

RUBE ADLER SPORTING GOODS   216-226-1740
11642 Detroit Avenue
Cleveland, Ohio 44102
   Discount to Nonprofit Organizations: Up to 20 percent. *Repairs:* Boots.

COLUMBUS

WILDERNESS TRACE, INC.　　　　　　614-457-8496
1295 Bethel Road
Columbus, Ohio 43220

Annual Sale: End of winter season—check with them for this year's date. *Discount:* 20 to 25 percent on a few items. *Discount to Nonprofit Organizations:* 5 to 10 percent. *Clearance Center:* 30 to 40 percent off list price on brands that may at times include Donner, Raichle, Lowa, Kelty, and Sierra Designs. *Repairs:* Some minor stitching repairs. *Special Features:* They run a kit-building clinic.

WILDERNESS TRACE, INC.　　　　　　614-294-5910
6 East 13th Street
Columbus, Ohio 43201

Annual Sale: End of winter season—check with them for this year's date. *Discount:* 20 to 25 percent on a few items. *Discount to Nonprofit Organizations:* 5 to 10 percent. *Clearance Center:* 30 to 40 percent off list price on brands that may at times include Donner, Raichle, Lowa, Kelty, and Sierra Designs. *Repairs:* Some minor repairs. *Special Features:* They run a kit-building clinic.

DAYTON

MENDELSON'S—THE STORE FOR ALL　　513-296-1222
OUTDOORS
2701 South Dixie Avenue
Dayton, Ohio 45409

Semiannual Sales: March for ski equipment, August for camping gear. *Manufacturers' Overruns:* 25 to 35 percent off list price; they carry a very good line of merchandise, so there should be good buys. *Repairs:* Skis.

WILDERNESS OUTFITTERS, INC.　　　513-252-5006
3962 Linden Avenue
Dayton, Ohio 45432

Discount: A limited range of merchandise is sold at less than list price. *Discount to Nonprofit Organizations:* 10 percent on orders over $100, 15 percent on orders over $500. *Manufacturers' Over-*

runs and Seconds: 10 to 20 percent off list price. *Rent:* Cross-country skis.

KENT

BACKWOODS TRADING COMPANY   216-673-0329
144 East Main Street, P.O. Box 717
Kent, Ohio 44240
    Discount to Nonprofit Organizations: Approximately 20 percent. *Rent:* Sleeping bags, tents, stoves, frame packs, soft packs, cross-country skis.

NORTH OLMSTEAD

RUBE ADLER   216-777-6550
25766 Lorain Road
North Olmstead, Ohio 44070
    Discount to Nonprofit Organizations: Up to 20 percent. *Repairs:* Boots.

SHEFFIELD LAKE

OHIO CANOE ADVENTURES   216-934-5345
5128 Colorado Avenue   Cleveland: 216-835-0861
Sheffield Lake, Ohio 44054
    Discount: 10 to 20 percent less than suggested list price on gear by North Face, Trailwise, Jansport, Ascente, Sierra Designs, Woolrich, Snow Lion, Class 5, Blacks, Alpine Designs, and many other top manufacturers; sometimes their discount is as high as 40 percent off list price—definitely worth the half-hour drive from Cleveland. *Discount to Nonprofit Organizations:* They give them. Ask. *Clearance Center:* They have one—worth checking every time you're near the store.

TERRACE PARK

WILDERNESS TRACE, INC.   513-831-3770
614 Wooster Park, Route 50
Terrace Park, Ohio 45174
    Discount: 20 to 25 percent discount on some items. *Clearance*

Center: 30 to 40 percent savings. *Special Features:* They run a kit-building clinic.

TOLEDO

CHURCHILL'S ADVENTURE SHOP   419-385-4599
2140 South Byrne Road
Toledo, Ohio 43614
   Annual Sale: February. *Discount to Nonprofit Organizations:* 15 percent. *Rent:* Tents, stoves, downhill and cross-country skis, kayaks, canoes. *Rental Sales:* End of each season—ask them about the dates of this year's sales. *R.P.*

THE MOUNTAIN MAN   419-893-9463
5353 Lewis
Toledo, Ohio 43612
   Rent: Tents, frame packs, soft packs, canoes, cross-country skis. *Rental Sale:* End of each season. *R.P.*

VIKING SHOP   419-537-0212
2735 North Reynolds
Toledo, Ohio 43615
   Repairs: Ski boots and tennis rackets, as well as packs. *Rent:* Tents, frame packs, soft packs, canoes, cross-country skis. *R.P.*

SYLVANIA

CHURCHILL'S ADVENTURE SHOPS   419-882-0051
5700 Monroe
Sylvania, Ohio 43560
   Rent: Tents, stoves, downhill and cross-country skis, kayaks, canoes. *Rental Sale:* End of each season. *R.P.*

_____ **Oklahoma**

OKLAHOMA CITY

BACKWOODS EQUIPMENT CO.   405-751-7376
10205 North May
Oklahoma City, Oklahoma 73120

Rent: Sleeping bags, foam pads, tents, frame packs, soft packs, kayaks, cross-country skis. *Rental Sale:* February. *R.P.*

BACKWOODS EQUIPMENT CO./
MOUNTAIN CHALET   405-239-7733
213 West Park Avenue
Oklahoma City, Oklahoma 73102
   Rent: Sleeping bags, foam pads, tents, frame packs, soft packs, kayaks, cross-country skis. *Rental Sale:* February. *R.P.*

THE WILDERNESS ADVENTURER   405-842-5447
5619 North Penn
Oklahoma City, Oklahoma
   Rent: Sleeping bags, foam pads, tents, frame packs, snowshoes, kayaks.

STILLWATER

THE WILDERNESS ADVENTURER   405-377-6004
Cowboy Mall
1118 Hall of Fame
Stillwater, Oklahoma 74074
   Rent: Sleeping bags, foam pads, tents, frame packs, snowshoes, kayaks.

TULSA

WILDERNESS ADVENTURER   918-663-4392
6508 East 51 Street
Tulsa, Oklahoma 74135
   Discount to Nonprofit Organizations: 10 percent to Scout troops only. *Rent:* Sleeping bags, foam pads, tents, frame packs, snowshoes, kayaks.

───────────────────────────────────── **Oregon**

ALBANY

HARVEY FOX'S SPORTHAUS   503-928-2244
222 Ellsworth
Albany, Oregon 97321

Manufacturers' Overruns and Seconds: 10 to 20 percent off list price. *Discount to Nonprofit Organizations:* 10 to 20 percent. *Clearance Center:* 20 percent less than list price. *Repairs:* Everything they sell. *Rent:* Frame packs, snowshoes, downhill and cross-country skis, ice axes, crampons, water skis, tennis rackets.

JERED'S OUTDOOR N' MORE   503-926-6067
2220 East Pacific Boulevard
Albany, Oregon 97321
    Manufacturers' Overruns and Seconds: 20 to 33 percent below regular price.

ASHLAND

THE SUN CYCLE COMPANY   503-482-5761
293 East Main Street
Ashland, Oregon 97520
    Discount: 10 percent to anyone purchasing $500 or more worth of merchandise. *Repairs:* Cross-country skis, downhill skis, bicycles. *Rent:* Cross-country skis.

BEND

JERED'S OUTDOOR N' MORE   503-382-8485
1500 North East 3rd
Bend, Oregon 97701
    Manufacturers' Overruns and Seconds: 20 to 33 percent below regular price.

CORVALLIS

HARVEY FOX'S SPORTHAUS   503-752-7299
137 Southwest 3rd
Corvallis, Oregon 97330
    Manufacturers' Overruns and Seconds: 10 to 20 percent less than list price. *Discount to Nonprofit Organizations:* 10 to 20 percent. *Clearance Center:* 20 percent discount on brands that may at times include Aslo boots, Fabiano, G.H. Bass, Hine/Snowbridge, Kelty, Alpenlite, Gerry, and White Stag gear. *Repairs:* Alpine and cross-country skis, guns, anything they sell. *Rent:* Frame packs, snowshoes, downhill and cross-country skis, ice axes, water skis, tennis rackets, crampons.

RECREATIONAL SPORTS WAREHOUSE, INC.   503-752-5612
311 Southwest Madison
Corvallis, Oregon 97330
   Rent: Tents, stoves, frame packs, soft packs, snowshoes, down-hill and cross-country skis, kayaks, ice axes, crampons. *Rental Sale:* At end of season. *R.P.*

EUGENE

BERG'S NORDIC SKI SHOP   503-343-0013
410 East 11th Street
Eugene, Oregon 97401
   Sales: They usually have three or four during the year. Check with them for this year's dates. *Repairs:* Downhill and cross-country skis. *Rent:* Sleeping bags, tents, hiking boots, frame packs, soft packs, snowshoes, downhill and cross-country skis, ice axes, crampons. *R.P.*

HARVEY FOX'S SPORTHAUS   503-342-7021
611 East 13th
Eugene, Oregon 97401
   Manufacturers' Overruns and Seconds: 10 to 20 percent less than usual list price. *Discount to Nonprofit Organizations:* 10 to 20 percent. *Clearance Center:* 20 percent discount. *Repairs:* Everything they sell. *Rent:* Frame packs, snowshoes, downhill and cross-country skis, crampons, water skis.

MATTOX OUTFITTERS   503-686-2332
57 West Broadway
Eugene, Oregon 97401
   Manufacturers' Overruns and Seconds: 10 to 60 percent less than list price. *Discount to Nonprofit Organizations:* 10 to 20 percent. *Clearance Center:* 20 to 50 percent discount on brands that may at times include Kelty, Hine/Snowbridge, Galibier. *Rent:* Cross-country skis.

MCMINNVILLE

JERED'S OUTDOOR N' MORE   503-472-4617
448 North 99 West
McMinnville, Oregon 97128

Semiannual Sales: February and July. *Discount:* They sell some gear at less than manufacturers' suggested list price. *Manufacturers' Overruns and Seconds:* 20 to 33 percent less than list price.

MILWAUKEE

JERED'S OUTDOOR N' MORE   503-659-3373
15100 S.E. McLoughlin Boulevard
Milwaukee, Oregon 97222
  Semiannual Sales: February and July. *Discount:* They sell some gear at less than manufacturers' suggested list price. *Manufacturers' Overruns and Seconds:* 20 to 33 percent less than list price.

OREGON CITY

JERED'S OUTDOOR N' MORE   503-565-0388
524 Main
Oregon City, Oregon 97045
  Semiannual Sales: February and July. *Discount:* They sell some gear at less than manufactures' suggested list price. *Manufacturers' Overruns and Seconds:* 20 to 33 percent less than list price.

PORTLAND

THE ADVANT EDGE   503-777-5521
324 N.E. 12th
Portland, Oregon 97232
  Semiannual Sales: March 1 and September 1. *Discount:* Some merchandise sold at less than manufacturers' suggested list price. *Discount to Nonprofit Organizations:* Up to 10 percent. *Manufacturers' Overruns and Seconds:* Up to 40 percent off list price. *Repairs:* Skis, daypacks, tennis rackets. *Rent:* Tents, boots, frame packs, soft packs, downhill and cross-country skis, ice axes, crampons. *Rental Sales:* March 1 and September 1. *R.P.*

ALPINE HUT, INC.   503-284-1164
1250 Lloyd Center
Portland, Oregon 97232
  Rent: Tents, boots, soft packs, ice axes, frame packs, crampons.

Beach Hiking
(Photo from the collection of North Country Mountaineering, Inc.)

THE GLACIER'S EDGE   503-255-7997
10108 N.E. Wasco Street
Portland, Oregon 97220

Semiannual Sales: March 1 and September 1. *Discount:* Some merchandise sold at less than manufacturers' suggested list price. *Discount to Nonprofit Organizations:* Up to 10 percent. *Manufacturers' Overruns and Seconds:* Up to 40 percent off list price. *Repairs:* Skis, daypacks, tennis rackets. *Rent:* Tents, hiking boots, frame packs, soft packs, downhill skis, ice axes, crampons, cross-country skis. *Rental Sales:* March 1 and September 1. *R.P.*

THE MOUNTAIN SHOP   503-288-6768
628 N.E. Broadway
Portland, Oregon 97232

Manufacturers' Seconds: They feature them occasionally at sales—20 to 30 percent less than usual price. *Discount to Nonprofit Organizations:* Depends upon the size of the order, and upon the organization. *Repairs:* Nordic and Alpine skis, stoves. *Rent:* Tents, boots, frame packs, snowshoes, downhill and cross-country skis, ice axes, crampons, tennis rackets. *R.P.*

THE MOUNTAIN SHOP PROGRESS   503-641-1991
8836 S.W. Hall Boulevard
Portland, Oregon 97223

Discount to Nonprofit Organizations: Depends upon the size of the order, and upon the organization. *Manufacturers' Seconds:* They feature them occasionally at sales—20 to 30 percent less than usual price. *Repairs:* Nordic and Alpine skis, stoves. *Rent:* Tents, boots, frame packs, downhill and cross-country skis, ice axes, crampons, tennis rackets. *R.P.*

SALEM

ANDERSON'S SPORTING GOODS   503-364-4400
340 Court Street
Salem, Oregon 97301

Annual Sale: Usually the end of September through October 1. *Manufacturers' Seconds:* Usually 20 percent off, sometimes more, depending on the item. *Clearance Center:* 20 to 40 percent off on brands that may at times include Kelty, North Face, Alpenlite, Camp 7, White Stag, Sport-Obermeyer boots, and Pivetta. *Repairs:* Skis, ski boots, minor stove repairs, scuba equipment. *Rent:* Frame packs, snowshoes, downhill and cross-country skis, ice axes, crampons, water skis, tennis rackets.

--- **Pennsylvania**

ALTOONA

THE PATHFINDER   814-943-4016
The Station Mall
Altoona, Pennsylvania 16603
   Discount: They sell brands like Camp Trails, Camp 7, Jansport, G.H. Bass, Fabiano, and Raichle at about 10 to 20 percent below suggested list price. *Discount to Nonprofit Organizations:* 10 to 20 percent. *Manufacturers' Seconds:* 15 to 25 percent below list price.

BRYN MAWR

JAMES L. COX SPORT SHOP   215-525-3163
931 Lancaster Avenue
Bryn Mawr, Pennsylvania 19010
   Discount to Nonprofit Organizations: They will sometimes give discounts of from 10 to 15 percent. *Repairs:* Backpacks, skis, tennis rackets. *Rent:* Tents, cross-country skis. *R.P.*

J.D. SACH'S WILDERNESS OUTFITTERS   215-527-3616
880 West Lancaster Avenue
Bryn Mawr, Pennsylvania 19010
   Discount to Nonprofit Organizations: It varies, depending upon size of order. Ask. *Manufacturers' Seconds:* They carry them but didn't tell us how much of a reduction they pass on to consumers. *Clearance Center:* 20 to 50 percent off list price on brands like Kelty, Galibier, Camp 7.

CAMP HILL

THE PATHFINDER   717-761-3906
Cedar Cliff Mall—1104 Carlisle Road
Camp Hill, Pennsylvania 17011
   Discount: They sell brands like Camp Trails, Jansport, G.H. Bass,

Fabiano, and Raichle at about 10 to 20 percent below suggested list price. *Discount to Nonprofit Organizations:* 10 to 20 percent. *Manufacturers' Seconds:* 15 to 25 percent below list price. *Rent:* Sleeping bags, tents, frame packs, snowshoes, canoes, cross-country skis. *Rental Sale:* February. *R.P.*

DOYLESTOWN

APPALACHIAN TRAIL OUTFITTERS   215-348-8069
Main & Oakland
Doylestown, Pennsylvania 18901
   Rent: Sleeping bags, foam pads, tents, stoves, frame packs, soft packs. *R.P.*

EASTON

FORKS VALLEY SPORTSWORLD   215-253-1111
P.O. Box 805
Easton, Pennsylvania 18042
   Rent: Sleeping bags, foam pads, parkas, tents, stoves, frame packs, soft packs, snowshoes, kayaks, canoes, ice axes, crampons, cross-country skis, lanterns. *Rental Sales:* End of season—ask them for this year's dates.

EXTON

WICK'S SKI SHOPS   215-363-1893
403 Pottstown Pike
Exton, Pennsylvania 19341
   Rent: Snowshoes, downhill and cross-country skis, canoes, kayaks. *R.P.*

GREENSBURG

EXKURSION   412-836-2703
530 South Main
Greensburg, Pennsylvania 15601
   Rent: Tents, frame packs, snowshoes. *R.P.*

HARRISBURG

THE PATHFINDER   717-761-3906
Cedar Cliff Mall
Harrisburg, Pennsylvania 17101
   Rent: Sleeping bags, tents, frame packs, snowshoes, canoes, cross-country skis. *Rental Sale:* February. *R.P.*

LANGHORNE

LANGHORNE SKI & SPORT   215-757-3113
2040 East Lincoln Highway
Langhorne, Pennsylvania 19047
   Discount to Nonprofit Organizations: 10 to 20 percent. *Manufacturers' Overruns and Seconds:* 30 to 40 percent less than list price. *Clearance Center:* 30 to 40 percent discount. *Repairs:* Alpine and cross-country skis. *Rent:* Sleeping bags, foam pads, tents, stoves, frame packs, soft packs, downhill and cross-country skis, crampons, canoes, ice axes.

OHIOPYLE

WILDERNESS VOYAGEURS OUTFITTERS   412-329-8336
Garrett Street—P.O. Box 97
Ohiopyle, Pennsylvania 15470
   Discount to Nonprofit Organizations: 15 percent. *Rent:* Canoes, cross-country skis.

PAOLI

JAMES L. COX SPORT SHOP   215-644-9325
23 Paoli Pike
Paoli, Pennsylvania 19301
   Discount to Nonprofit Organizations: 10 to 15 percent, but not given to all organizations; ask them. *Repairs:* Backpacks, skis, tennis rackets. *Rent:* Tents, cross-country and downhill skis. *R.P.*

PHILADELPHIA

BASE CAMP   215-563-9626
1730 Chestnut Street
Philadelphia, Pennsylvania 19103
    Manufacturers' Seconds: They sometimes carry seconds of Caribou tents at 15 percent discount. *Rent:* Tents, cross-country skis. *Rental Sale:* End of season—ask them for this year's dates.

I. GOLDBERG COMPANY   215-925-9393
902 Chestnut Street
Philadelphia, Pennsylvania 19107
    Discount to Nonprofit Organizations: 5 to 10 percent. *Manufacturers' Overruns and Seconds:* 20 to 30 percent less than list price. *Repairs:* Tents, sleeping bags, zippers.

PITTSBURGH

BONN'S OUTDOOR ARMY-NAVY
SURPLUS, INC.   412-881-4774
3300 Saw Mill Run Boulevard (Rt. 51 South)
Pittsburgh, Pennsylvania 15227
    Discount: 10 to 20 percent less than list price on merchandise by Coleman, Eastern Canvas, Henderson, I. Spiewak & Sons. *Discount to Religious Organizations:* 10 percent.

THE MOUNTAIN TRAIL SHOP   412-687-1700
5435 Walnut Street
Pittsburgh, Pennsylvania 15232
    Discount to Nonprofit Organizations: 20 percent. *Clearance Center:* 20 to 80 percent reductions. *Repairs:* Stoves, boots, packs, etc. *Rent:* Sleeping bags, foam pads, tents, frame packs, cross-country skis. *Rental Sales:* End of season. *R.P.*

SPRINGFIELD

WICK'S SKI SHOPS, INC.   215-543-5445
321 West Woodland Avenue
Springfield, Pennsylvania 19064
    Rent: Snowshoes, downhill and cross-country skis. *R.P.*

STATE COLLEGE

THE PATHFINDER   814-237-8086
137 East Beaver Street
State College, Pennsylvania 16801
Annual Sale: February. *Discount:* Some merchandise sold at less than list price. *Discount to Nonprofit Organizations:* It varies. Ask them. *Manufacturers' Seconds:* At least 15 percent discount. *Clearance Center:* 20 to 50 percent reductions. *Rent:* Sleeping bags, tents, frame packs, snowshoes, canoes, cross-country skis. *Rental Sale:* February. *R.P.*

WHITEHALL

WILDERNESS TRAVEL OUTFITTERS   215-439-9754
2530 MacArthur Road
Whitehall, Pennsylvania 18052
Rent: Snowshoes, cross-country skis.

WILKES-BARRE

UNCLE EYEBALL'S MOUNTAIN
TRAVELERS' EMPORIUM   717-829-1409
35 East South Street
Wilkes-Barre, Pennsylvania 18702
Discount to Nonprofit Organizations: 20 percent. *Manufacturers' Overruns and Seconds:* 10 percent off list price. *Rent:* Foam pads, tents, stoves, frame packs, snowshoes, kayaks, canoes, ice axes, crampons, cross-country skis.

YARDLEY

J. D. SACH'S WILDERNESS OUTFITTERS   215-493-4536
10 Penn Valley Drive
Yardley, Pennsylvania 19067
Discount to Nonprofit Organizations: It varies, depending upon quantity. Ask. *Clearance Center:* 20 to 50 percent off list price on brands that will occasionally include Kelty, Camp 7, Galibier, and Snow Lion. *Rent:* Tents, kayaks, cross-country skis.

CRANSTON

THE OUTDOORSMAN   401-943-2271
753 Oaklawn Avenue
Cranston, Rhode Island 02920
   Discount to Nonprofit Organizations: 10 to 20 percent. *Manufacturers' Seconds:* They have them but didn't tell us how much of a reduction they pass on to customers. Ask. *Rent:* Sleeping bags, tents, stoves, frame packs, snowshoes, canoes, ice axes, cross-country skis. *Rental Sale:* April. *R.P.*

EAST GREENWICH

FIN & FEATHER LODGE   401-884-4432
95 Frenchtown Road—Route 402
East Greenwich, Rhode Island 02818
   Discount: 20 percent below suggested list price on brands like Sorrel, Orvis, Dunham, Camp Trails, Woods, and Woolrich; this is one of *very* few places in the United States where you can find discounts on Orvis products. *Rental Sale:* End of summer season— ask them for the exact date. *R.P.*

PROVIDENCE

THE OUTDOORSMAN   401-274-6770
110 Waterman Street
Providence, Rhode Island 02906
   Annual Sale: April. *Discount to Nonprofit Organizations:* 10 to 20 percent. *Manufacturers' Seconds:* 20 percent discount. *Repairs:* Stoves. *Rent:* Sleeping bags, tents, stoves, frame packs, soft packs, canoes, ice axes, cross-country skis. *Rental Sales:* April. *R.P.*

THE SUMMIT SHOP   401-751-5052
185 Wayland Avenue
Providence, Rhode Island 02906
   Semiannual Sales: October 1 and late February. *Discount to Nonprofit Organizations:* Approximately 20 percent. *Manufacturers' Seconds:* At least 25 percent discount. *Clearance Center:* 25 to 30

percent off list price on brands that will at times include Galibier, Fabiano, Camp 7, Kelty, and Mountain Equipment. *Rent:* Foam pads, tents, stoves, frame packs, snowshoes, kayaks, ice axes, cross-country skis. *R.P.*

WAKEFIELD

THE OUTDOORSMAN   401-783-8515
415 Kingstown Road
Wakefield, Rhode Island 02879
    Annual Sale: April. *Discount to Nonprofit Organizations:* 10 to 20 percent. *Manufacturers' Seconds:* 20 percent. *Repairs:* Stoves. *Rent:* Sleeping bags, tents, stoves, frame packs, snowshoes, canoes, ice axes, cross-country skis. *Rental Sale:* April. *R.P.*

―――――――――――――――――――――― **South Carolina**

CLEMSON

THE GOOD EARTH   803-654-1325
University Square Mini Mall
Clemson, South Carolina 29631
    Discount to Nonprofit Organizations: 10 percent. *Clearance Center:* 10 to 40 percent discount on brands that will occasionally include Vasque, Gerry, Trailwise, and Wilderness Experience. *Rent:* Sleeping bags, foam pads, tents, stoves, frame packs, kayaks, canoes, ice axes, cross-country skis. *Special Features:* They offer instruction in caving.

―――――――――――――――――――――**South Dakota**

MITCHELL

GRIGGS SPORT CENTER   605-996-6686
Super City Shopping Mall
1801 North Main
Mitchell, South Dakota 57301
    Discount: They sell some gear at less than list price. *Discount to*

Nonprofit Organizations: 10 to 23 percent. *Repairs:* Stoves, lanterns, etc. *Rent:* Cross-country skis.

RAPID CITY

DU-ELL SPORTING GOODS   605-343-7345
732 Jackson Boulevard
Rapid City, South Dakota 57701
  Annual Sale: Around January 15. *Discount to Nonprofit Organizations:* 15 to 20 percent.

HOMESPUN   605-348-6427
609 Mount Rushmore Road
Rapid City, South Dakota 57701
  Annual Sale: Their parkas and vests go on sale in February. *Discount to Nonprofit Organizations:* 10 percent. *Repairs:* They do repairs on garments or bags made from Frostline kits. *Special Features:* They will customize kits on a limited basis. They'll also construct some kits for customers.

MOUNTAIN GOAT SPORTS   605-342-4165
2111 Jackson Boulevard
Rapid City, South Dakota 57701
  Semiannual Sales: Fall and spring—check with them for this year's dates. *Discount:* A few items are sold at less than suggested list price. *Discount to Nonprofit Organizations:* Depends upon quantity purchased. *Repairs:* Ski equipment.

SIOUX FALLS

SUN & FUN SPECIALITY SPORTS   605-338-1351
Western Mall
Sioux Falls, South Dakota 57104
  Rent: Downhill and cross-country skis. *R.P.*

CHATTANOOGA

CHATTANOOGA OUTDOORS   615-899-2525
6901 Lee Highway
Chattanooga, Tennessee 37421
   Semiannual Sales: One winter sale, one summer sale—check
with them for this year's dates. *Discount to Nonprofit Organizations:* It varies. *Manufacturers' Overruns and Seconds:* 20 to 30
percent discount. *Clearance Center:* 20 to 30 percent discount on
brands that will upon occasion include Fabiano, Alpine Designs,
Jansport, Woolrich, Snow Lion. *Rent:* Sleeping bags, tents, frame
packs, soft packs, snowshoes, downhill skis. *R.P.*

CHATTANOOGA OUTDOORS   615-870-2344
Highland Plaza
Chattanooga, Tennessee 37415
   Semiannual Sales: One winter sale, one summer sale—check
with them for this year's dates. *Discount to Nonprofit Organizations:* It varies. *Manufacturers' Overruns and Seconds:* 20 to 30
percent discount. *Clearance Center:* 20 to 30 percent discount on
brands that will upon occasion include Fabiano, Alpine Designs,
Jansport, Woolrich, Snow Lion.

GATLINBURG

FAMILY TRAILS   615-436-7979
964 River Road
Gatlinburg, Tennessee 37738
   Semiannual Sales: One winter sale, one summer sale—check
with them for this year's dates. *Discount to Nonprofit Organizations:* It varies. *Manufacturers' Overruns and Seconds:* 20 to 30
percent discount. *Clearance Center:* 20 to 30 percent discount on
brands that will upon occasion include Fabiano, Alpine Designs,
Jansport, Woolrich, and Snow Lion. *Rent:* Sleeping bags, tents,
frame packs, soft packs, snowshoes, downhill skis. *R.P.*

KNOXVILLE

BLUE RIDGE MOUNTAIN SPORTS LTD.   615-584-9432
5201 Kingston Pike
Knoxville, Tennessee 37919
   Semiannual Sales: January and June or July. *Discount to Non-profit Organizations:* 10 percent. *Clearance Center:* 30 percent discount. *Rent:* Tents, stoves, frame packs, kayaks, life vests. *R.P.*

OUTDOOR AMERICA OUTFITTERS   615-693-5262
114 Northshore Drive
Knoxville, Tennessee 37912
   Rent: Sleeping bags, foam pads, tents, frame packs, snowshoes. *R.P.*

MEMPHIS

THE CAMP & HIKE SHOP   901-365-4511
4674 Knight Arnold Road
Memphis, Tennessee 38118
   Discount to Nonprofit Organizations: 10 to 15 percent. *Manufacturers' Overruns and Seconds:* When available, they offer them at from 20 to 40 percent off list price. *Clearance Center:* 20 to 40 percent off list price, on brands that will upon occasion include Alpine Designs, Camp Ways, Coleman, Dunham. *Repairs:* All stoves, lanterns, tents. *Rent:* Tents, stoves, frame packs.

CAMPER'S CORNER   901-946-2566
2050 Elvis Presley Boulevard
Memphis, Tennessee 38106
   Semiannual Sales: March and October. *Discount to Nonprofit Organizations:* 15 percent on bulk orders. *Manufacturers' Overruns:* 10 to 15 percent off list price. *Clearance Center:* 20 to 50 percent on brands that will occasionally include North Face, Alpine Designs, Dunham, Fabiano. *Repairs:* Skis.

ABILENE

WILDERNESS TRAIL EQUIPMENT COMPANY   915-677-5746
3005 South First Street
Abilene, Texas 79605
 Semiannual Sales: August and September. *Discount:* 10 percent
or more off manufacturers' suggested list price on brands like Camp
Trails, White Stag, Himalayan Industries backpacks, and Dexter
boots. *Discount to Nonprofit Organizations:* Depends upon size of
order. *Clearance Center:* Reductions of from 20 to 30 percent.
Repairs: Tents, packs, frames, stoves, etc.

ARLINGTON

SOUTHWEST CANOE N KAYAK OUTFITTERS   817-461-4503
2002 West Pioneer Parkway
Arlington, Texas 76013
 Semiannual Sales: February and September. *Discount to Nonprofit Organizations:* Sometimes. *Clearance Center:* 20 percent reductions. *Rent:* Sleeping bags, tents, frame packs, soft packs,
kayaks, canoes. *Rental Sales:* February and September. *R.P.*

AUSTIN

WHOLE EARTH PROVISION COMPANY   512-478-1577
2410 San Antonio Street
Austin, Texas 78705
 Manufacturers' Seconds: 10 percent discount, when they can
get them.

DALLAS

THE MOUNTAIN CHALET   214-363-0372
5500 Greenville Avenue
Dallas, Texas 75206
 Semiannual Sales: Spring and fall—check with them for this
year's dates. *Rent:* Sleeping bags, foam pads, tents, frame packs,
soft packs, kayaks, cross-country skis. *Rental Sale:* February.
R.P.

MOUNTAIN HIDEOUT 214-234-8651
14010 Coit Road
Dallas, Texas 75240
   Discount to Nonprofit Organizations: 15 percent.

DENTON

WILDERNESS OUTFITTERS, INC. 817-387-5900
1512 West Hickory
Denton, Texas 75201
   Annual Sale: Late winter—check for this year's date. *Discount to Nonprofit Organizations:* 10 percent. *Clearance Center:* 30 percent discount.

FORT WORTH

WILDERNESS OUTFITTERS, INC. 817-332-2423
3212 Camp Bowie
Fort Worth, Texas 76107
   Annual Sale: Late winter—Check with them for this year's date. *Discount to Nonprofit Organizations:* 10 percent. *Clearance Center:* 30 percent discount. *Rent:* Sleeping bags, foam pads, stoves, frame packs, soft packs, kayaks, canoes, ice axes, crampons. *R.P.*

LUBBOCK

THE SPORT HAUS 806-747-1681
2309 Broadway
Lubbock, Texas 79401
   Annual Sale: March. *Discount to Nonprofit Organizations:* 10 percent. *Clearance Center:* 30 to 40 percent discount on brands that will upon occasion include Jansport, White Stag, Alpine Designs, Kelty, Raichle, and Lowa boots. *Repairs:* Skis.

SAN ANTONIO

MURSCH SPORTS HEADQUARTERS 512-684-2020
6132 Bandera
San Antonio, Texas 78238
   Semiannual Sales: April and November. *Discount to Nonprofit*

Organizations: 10 percent. *Clearance Center:* Reductions of from 33 to 50 percent.

Utah

SALT LAKE CITY

FOOTHILL VILLAGE SPORTS DEN   801-582-5611
1384 Foothill Drive
Salt Lake City, Utah 84108
   Semiannual Sales: August 10 and February 18. *Discount to Nonprofit Organizations:* 10 to 20 percent. *Repairs:* Boots, skis, etc. *Rent:* Sleeping bags, foam pads, tents, stoves, frame packs, snowshoes, downhill and cross-country skis.

HOLUBAR STORES   801-621-3071
4385 South State Street
Salt Lake City, Utah 84107
   Semiannual Sales: Spring and end of summer—check with them for this year's dates. *Discount:* A few items are sold at less than list price. *Discount to Nonprofit Organizations:* Check with them; they do give them. *Manufacturers' Seconds:* They get a few of their own Holubar seconds at from 10 to 50 percent less than list price. *Repairs:* Cross-country ski equipment, Optimus stoves, Holubar label merchandise. *Rent:* Sleeping bags, foam pads, tents, frame packs, soft packs, snowshoes, ice axes, cross-country skis.

MOUNTAINEER SPORTS   801-582-2338
155 Trolley Square
Salt Lake City, Utah 84102
   Discount to Nonprofit Organizations: A generous 20 percent. *Manufacturers' Overruns and Seconds:* 10 percent less than list price. *Repairs:* All items sold. *Rent:* Sleeping bags, foam pads, tents, stoves, frame packs, soft packs, snowshoes, downhill skis, kayaks. *R.P.*

MOUNTAINEER SPORTS   801-582-0427
207 South 13th East
Salt Lake City, Utah 84102
   Discount to Nonprofit Organizations: A generous 20 percent.

Manufacturers' Overruns and Seconds: 10 percent less than list price. *Repairs:* All items sold. *Rent:* Sleeping bags, foam pads, tents, stoves, frame packs, soft packs, snowshoes, downhill skis, kayaks, tennis rackets. *R.P.*

TIMBERLINE SPORTS, INC.   801-466-2101
3155 South Highland Drive
Salt Lake City, Utah 84106

Annual Sale: March. *Discount to Nonprofit Organizations:* 20 percent—you're lucky! Salt Lake City stores give some of the largest discounts to nonprofit organizations we've found anyplace in the United States. *Clearance Center:* 20 to 30 percent discount on brands that will upon occasion include Hine/Snowbridge, North Face, Kelty, Wilderness Experience, Pivetta, and Tecnica boots. Watch this clearance department; they carry excellent merchandise. *Repairs:* All backpacking-related repairs, including cross-country skis and kayak repairs. *Rent:* Sleeping bags, foam pads, tents, stoves, frame packs, soft packs, snowshoes, kayaks, ice axes, crampons, cross-country skis.

Vermont

BENNINGTON

VERMONT COUNTRY SPORTS, INC. 802-442-5530
201 South Street
Bennington, Vermont 05201

Semiannual Sales: March and July. *Discount to Nonprofit Organizations:* 10 percent. *Manufacturers' Overruns:* 30 to 40 percent below list price—a very good discount. *Clearance Center:* 30 to 40 percent savings on brands that may upon occasion include Gerry, Raichle, and Wilderness Experience. *Repairs:* Skis, tents, canoes. *Rent:* Tents, frame packs, soft packs, snowshoes, downhill skis, canoes. *Rental Sales:* March and July.

BURLINGTON

DAKINS MOUNTAIN SHOP   802-863-5581
227 Main Street
Burlington, Vermont 05401
   Discount to Nonprofit Organizations: It varies. Ask them.
Rent: Sleeping bags, tents, snowshoes, cross-country skis. *R.P.*

ESSEX JUNCTION

ESSEX RENTAL & SALES CENTER   802-878-5316
26 Kellogg Road—P.O. Box 359
Essex Junction, Vermont 05452
   Rent: Tents, stoves, frame packs, soft packs, snowshoes, canoes.

MANCHESTER

THE VALLEY VIKING   802-362-1661
Route 7
Manchester, Vermont 05354
   Annual Sale: Washington's Birthday. *Repairs:* Cross-country skis.
Special Features: They have an interesting range of imported cross-
country ski clothing and accessories.

MIDDLEBURY

SKIHAUS MOUNTAIN SPORTS   802-388-2823
Box 509—Merchants Square
Middlebury, Vermont 05753
   Discount to Nonprofit Organizations: 20 percent. *Manufactur-
ers' Overruns:* Occasionally—worth checking. *Repairs:* Skis, tennis
rackets, bicycles. *Rent:* Frame packs, downhill and cross-country
skis, bicycles. *Rental Sale:* End of season—ask them for the dates.
R.P.

MONTPELIER

ONION RIVER SPORTS, INC.   802-229-9409
20 Langdon Street
Montpelier, Vermont 05602
   Discount: They are one of very few discounters in the state. Their

prices are generally 10 to 15 percent off list price on brands like Gerry, Tecnica boots, and Ascente sleeping bags. *Discount to Nonprofit Organizations:* They give a very good discount, 20 to 30 percent; even EMS gives only 5 to 15 percent. *Repairs:* Everything they sell. *Rent:* Snowshoes, bicycles, cross-country skis. *R.P.*

ST. ALBANS

SPORT SHOP, INC.   802-524-3312
139 Lake Street
St. Albans, Vermont 05478   •
   Semiannual Sales: Spring and fall—check with them for dates. *Discount:* An average of 4 percent below list price on brands like Alpine Designs, Climb High backpacks, Hunter sleeping bags, and Greylock Mountain Industries sleeping bags. *Repairs:* Ski equipment.

SHELBURNE

CANOE IMPORTS, INC.   802-985-2992
R.D. 2—Box 1. Route 7
Shelburne, Vermont 05482
   Annual Sale: Each fall—check with them for this year's date. *Discount to Nonprofit Organizations:* 10 percent. *Manufacturers' Seconds:* 10 to 25 percent below list price. *Repairs:* Canoes and small craft.

SOUTH BURLINGTON

EASTERN MOUNTAIN SPORTS   802-864-0473
100 Dorset Street—Box 2142
South Burlington, Vermont 05401
   Sales: Early October, December, April. *Discount:* Their own brand of merchandise represents a savings of from 10 to 20 percent on the cost of comparable brand name products. *Discount to Nonprofit Organizations:* It varies—only on bulk orders. *Manufacturers' Overruns and Seconds:* 20 to 25 percent savings on these. *Repairs:* Some minor repairs. *Rent:* Ice axes, crampons, Mickey Mouse boots, tents, foam pads, stoves, frame packs, soft packs, cross-country skis. *R.P.*

BRISTOL

TREK MOUNTAIN SPORTS   703-669-1213
1010 Commonwealth
Bristol, Virginia 37620
   Semiannual Sales: March and August. *Discount:* They sometimes sell merchandise at 10 percent less than list price. *Discount to Nonprofit Organizations:* 5 to 25 percent. *Clearance Center:* 25 percent savings on brands that will occasionally include Alpenlite, Camp Trails, Coleman, and Kelty. *Repairs:* Rafts, packs, tents, bicycles. *Rent:* Sleeping bags, foam pads, tents, stoves, frame packs, soft packs, snowshoes, kayaks, canoes, cross-country skis. *R.P.*

CHARLOTTESVILLE

BLUE RIDGE MOUNTAIN SPORTS   804-977-4400
1417 Emmet Street
Charlottesville, Virginia 22901
   Annual Sale: June or July. *Discount to Nonprofit Organizations:* It varies with purchase and group. *Clearance Center:* 30 percent reductions. *Repairs:* Stoves. *Rent:* Tents, stoves, frame packs, kayaks, life vests. *R.P.*

LYNCHBURG

BLUE RIDGE MOUNTAIN SPORTS   804-384-9113
Forest Hills Shopping Center, Linkhorne Drive
Lynchburg, Virginia 24501
   Annual Sale: June or July. *Discount to Nonprofit Organizations:* It varies with purchase and group. *Clearance Center:* 30 percent reductions. *Repairs:* Stoves. *Rent:* Tents, stoves, frame packs, kayaks, canoes.

NORFOLK

BLUE RIDGE MOUNTAIN SPORTS   804-461-2767
881 North Military Highway
Norfolk, Virginia 23502
   Semiannual Sales: January, and June or July. *Discount to Non-*

profit Organizations: 10 percent. *Clearance Center:* 30 percent saving on brands that may occasionally include Kelty, North Face, Fabiano. *Rent:* Tents, stoves, frame packs, kayaks, canoes, life vests. *Rental Sale:* Every year—ask them for this year's date. *R.P.*

OAKTON

APPALACHIAN OUTFITTERS   703-281-4324
2938 Chain Bridge Road—Box 249
Oakton, Virginia 22124

Discount: According to their own figures, they sell 29 or 30 items cheaper than at the nearby Dart Drug Discount stores. *Discount to Nonprofit Organizations:* It varies. Ask them. *Manufacturers' Overruns and Seconds:* They have them occasionally, at from 5 to 10 percent less than usual list price. *Repairs:* Any sewing repairs on nylon. *Rent:* Sleeping bags, foam pads, stoves, frame packs, hiking boots, cross-country skis, canoes. *R.P.*

RICHMOND

ALPINE OUTFITTERS, INC.   804-794-4172
11010 Midlothian Turnpike
Richmond, Virginia 23235

Semiannual Sales: March and August. *Discount to Nonprofit Organizations:* 10 percent on Eureka and Camp Trails. *Manufacturers' Overruns and Seconds:* They have them sometimes at 10 to 20 percent below list price. *Repairs:* Stoves, skis. *Rent:* Sleeping bags, foam pads, tents, stoves, frame packs, soft packs, kayaks, canoes. *Rental Sales:* September or October. *R.P.:* Sometimes.

SALEM

APPALACHIAN OUTFITTERS   703-389-1056
Old Logans Barn
Route 460 South
Salem, Virginia 24153

Rent: Sleeping bags, foam pads, stoves, frame packs, hiking shoes, cross-country skis, canoes. *Rental Sale:* First week of December. *R.P.*

SPRINGFIELD

HERMAN'S WORLD OF SPORTING GOODS   703-971-8303
6787 Springfield Mall Shopping Center
Springfield, Virginia 22150
   Annual Sale: Around November 28. *Discount:* 5 to 10 percent discount on some merchandise, including Coleman products.

VIENNA

HERMAN'S WORLD OF SPORTING GOODS   703-790-9810
8204 Leesburg Pike
Tyson's Corner
Vienna, Virginia 22101
   Annual Sale: Around November 28. *Discount:* 5 to 10 percent discount on some merchandise, including Coleman products.

-- **Washington**

FEDERAL WAY

ALPINE HUT   206-522-7787
32015 23rd South
Federal Way, Washington 98003
   Semiannual Sales: August and February. *Discount to Nonprofit Organizations:* 10 to 15 percent. *Manufacturers' Overruns and Seconds:* 25 to 50 percent discount. *Rent:* Tents, hiking and climbing boots, frame packs, snowshoes, downhill skis, ice axes, crampons. *R.P.*

LEAVENWORTH

DER SPORTSMANN   509-548-5623
837 Front Street
Leavenworth, Washington 98826
   Annual Sale: March. *Discount:* They sell some gear at less than manufacturers' list price "only to be competitive." *Manufacturers'*

Camp Muir, Mount Rainier
(Photo from the collection of North Country Mountaineering, Inc.)

Overruns and Seconds: 30 percent less than list price. *Clearance Center:* 40 percent off on brands that may include Kelty, Sierra Designs, Camp Trails, Fabiano, or several other top brands. *Rent:* Tents, stoves, frame packs, soft packs, snowshoes, ice axes, crampons, cross-country skis. *R.P.*

PULLMAN

NORTHWESTERN MOUNTAIN SPORTS   509-567-3981
North 115 Grand
Pullman, Washington 99163
    Annual Sale: End of the ski season. *Discount:* Most of their merchandise is sold at manufacturers' suggested markup; a few items are discounted 15 to 20 percent, which means you ought to study the mail order catalogs or in some other way keep yourself current on what the things you want cost. That way you won't buy when you know you can get something cheaper. You'll also be able to recognize a bargain when you see one. *Repairs:* Ski equipment. *Rent:* Cross-country skis.

SEATTLE

EDDIE BAUER EXPEDITION OUTFITTER   206-622-2766
1926 3rd Avenue
Seattle, Washington 98101
    Manufacturers' Seconds: Here's a treasure trove if you have a yen to own some of the finest gear on the market: Eddie Bauer's own brand. If there are Eddie Bauer seconds (and there are) this is where they'll turn up: at the relatively new Bauer seconds and discontinued shop, The Loft. Discounts here are from 30 to 50 percent. You'll find other good brands here too. *Repairs:* They repair some footwear and dry clean some down products. They also do some tailoring. *Rent:* Tents, frame packs. *Rental Sales:* July and January. *R.P.*

FIORINI'S SPORTS   206-523-9610
4720 University Village Place
Seattle, Washington 98105

Manufacturers' Overruns and Seconds: 30 to 50 percent less than list price.

MOUNTAIN SAFETY RESEARCH STORE   206-324-5730
110 East Pike Street
Seattle, Washington 98122
    Semiannual Sales: First week in April, and July or August. *Discount to Nonprofit Organizations:* 4 percent on mail order sales over $400. *Repairs:* They repair MSR-manufactured items. *Rent:* Snowshoes, ice axes, crampons, cross-country skis. *R.P.*

THE NORTH FACE   206-633-4431
4560 University Way, N.E.
Seattle, Washington 98105
    Semiannual Sales: April and October. *Manufacturers' Seconds:* If you're a fan of North Face equipment (and who isn't?), you'll find this an excellent place to watch for North Face seconds at 20 percent discount. *Repairs:* All North Face equipment. *Rent:* Foam pads, tents, frame packs, soft packs, ice axes, crampons, cross-country skis. *Rental Sales:* April and October. *R.P.*

SPOKANE

SELKIRK BERGSPORT   509-328-5020
West 30 International Way—Box 2701
Spokane, Washington 99220
    Discount to Nonprofit Organizations: 15 percent. *Repairs:* Stoves. *Rent:* Tents, frame packs, snowshoes, ice axes, crampons, cross-country skis. *Rental Sale:* March 1. *R.P.*

WESTERN OUTDOOR SPORTS   509-926-1543
North 111 Vista
Spokane, Washington 99206
    Semiannual Sales: March and October. *Discount to Nonprofit Organizations:* 10 to 15 percent. *Repairs:* Clothing, packs, tents, etc. *Rent:* Sleeping bags, tents, frame packs, snowshoes, ice axes, crampons, cross-country skis. *R.P.*

Along the West Coast Trail
(Photo from the collection of North Country Mountaineering, Inc.)

TACOMA

B & I MOUNTAINEERING   206-584-3207
8012 South Tacoma Way
Tacoma, Washington 98499

Discount to Nonprofit Organizations: 10 to 15 percent. *Manufacturers' Overruns:* 20 to 30 percent discount. *Repairs:* Coleman, Optimus, Primus. *Rent:* Sleeping bags, foam pads, tents, hiking boots, climbing boots, stoves, frame packs, soft packs, snowshoes, canoes, ice axes, crampons, cross-country skis. *R.P.*

ASPLUND'S SKI TOURING & MOUNTAINEERING

509-662-6539

1544 North Wenatchee Avenue
Wenatchee, Washington 98801

Semiannual Sales: March and September. *Discount to Nonprofit Organizations:* 10 percent or more, depending upon items ordered and the amount. *Repairs:* All items sold, except repairs requiring a sewing machine. *Rent:* Tents, frame packs, soft packs, snowshoes, ice axes, crampons, cross-country skis. *Rental Sales:* September and March. *R.P.:* Some items; only the first rental fee applies. *Special Features:* During the winter season they give free cross-country ski lessons on Saturdays and Sundays. Ask Janice Asplund at the store about them.

———————————————————— **West Virginia**

HUNTINGTON

THE CAMP SITE

304-525-6471

1518 Fourth Avenue
Huntington, West Virginia 25701

Discount to Nonprofit Organizations: They give 10 percent only on items ordered over $200; no discount on stocked merchandise. *Rentals:* Tents, stoves, frame packs, soft packs, canoes. *R.P.*

MORGANTOWN

THE PATHFINDER OF WEST VIRGINIA, LTD.   304-292-5223

182 Willey Street
Morgantown, West Virginia 26505

Discount: They sometimes sell merchandise at less than suggested retail price. *Discount to Nonprofit Organizations:* 10 percent or more, depending on the item. *Manufacturers' Overruns and Seconds:* They occasionally have them but didn't specify the size of the discount. *Repairs:* Bicycles, skis, tents, sleeping bags, etc. *Rent:* Sleeping bags, foam pads, tents, stoves, frame packs, downhill and cross-country skis, canoes. *R.P.*

MOUTH OF SENECA

THE GENDARME
P.O. Box 53
Mouth of Seneca, West Virginia 26884

Semiannual Sales: Spring and fall—don't be afraid to ask when this year's sales begin. *Discount:* They charge 10 percent less than manufacturers' list price on brands like Alpine Designs, Trailwise, Galibier. John Markwell also carries technical climbing equipment and caving supplies. Considering what a good climbing rope costs today, 10 percent savings will pay for dinner plus a glass of wine. *Discount to Nonprofit Organizations:* 10 percent.

Wisconsin

BRULE

BRULE RIVER CANOE RENTAL   715-372-4983
Box 90
Brule, Wisconsin 54820

Discount: They offer a discount on kayaks and canoes but didn't tell us how much. *Manufacturers' Seconds:* Canoes. *Repairs:* Canoes. *Rent:* Kayaks and canoes.

KENOSHA

THE PACK SHOP   414-654-3351
5033 6th Avenue
Kenosha, Wisconsin 53140

Annual Sale: Each spring—ask them for this year's date. *Discount to Nonprofit Organizations:* 10 percent. *Clearance Center:* 10 to 20 percent off list price on brands that may include Gerry, Camp 7, North Face, Galibier. *Repairs:* Stoves, skis, kayaks. *Rent:* Tents, frame packs, soft packs, kayaks, cross-country skis. *R.P.*

MADISON

EREWHON MOUNTAIN SUPPLY   608-251-9059
401 State Street
Madison, Wisconsin 53703

Semiannual Sale: March and early September. *Discount:* They sometimes sell merchandise at less than list price—"usually *not*, unless competition warrants." *Discount to Nonprofit Organizations:* 5 to 20 percent, and special rental rates for nonprofit groups. *Repairs:* Cross-country skis, stoves, some tents and other soft goods. *Rent:* Sleeping bags, foam pads, tents, frame packs, snowshoes, cross-country skis.

H. H. PETRIE SPORTING GOODS, INC.   608-231-2441
702 North Midvale
Madison, Wisconsin 53705

Annual Sale: Labor Day weekend ski sale. *Discount to Nonprofit Organizations:* 10 percent. *Manufacturers' Overruns and Seconds:* 20 to 50 percent discount. *Repairs:* Everything but sewing repairs. *Rent:* Cross-country skis, downhill skis. *Rental Sale:* Labor Day. *R.P.:* Within 10 days of the rental.

H. H. PETRIE SPORTING GOODS, INC.   608-256-0814
1406 Emil Street
Madison, Wisconsin 53705

Annual Sale: Labor Day weekend ski sale. *Discount to Nonprofit Organizations:* 10 percent. *Manufacturers' Overruns and Seconds:* 20 to 50 percent discount. *Repairs:* Everything but sewing repairs. *Rent:* Cross-country skis, downhill skis. *Rental Sale:* Labor Day. *R.P.:* Within 10 days of the rental.

H. H. PETRIE SPORTING GOODS, INC.   608-256-1347
644 State Street
Madison, Wisconsin 53703

Annual Sale: Labor Day weekend ski sale. *Discount to Nonprofit Organizations:* 10 percent. *Manufacturers' Overruns and Seconds:* 20 to 50 percent discount. *Repairs:* Everything but sewing repairs. *Rent:* Cross-country skis, downhill skis. *Rental Sale:* Labor Day. *R.P.:* Within 10 days of the rental.

MILWAUKEE

LAACKE & JOYS, INC.   414-271-7878
1433 North Water Street
Milwaukee, Wisconsin 53202

Annual Sale: Last weekend in August. *Discount:* On a few items. *Discount to Nonprofit Organizations:* It varies, but they do give them. Ask. *Repairs:* Skis. *Rent:* Downhill and cross-country skis.

PETRIE'S SEA 'N SKI   414-462-5880
4248 North 76 Street
Milwaukee, Wisconsin 53222

Annual Sale: August. *Manufacturers' Overruns and Seconds:* 30 to 50 percent discount.

UNITED MILITARY SUPPLY   414-272-3574
533 West Wisconsin Avenue
Milwaukee, Wisconsin 53203

Discount: They sell brands like Optimus, White Stag, Himalayan, and World Famous at an average 10 percent discount. They also handle United States military surplus gear. *Discount to Nonprofit Organizations:* 10 percent. *Manufacturers' Overruns and Seconds:* 25 percent discount.

OSHKOSH

HIKE OUT, LTD.   414-235-2720
2189 Abraham Lane—P.O. Box 374
Oshkosh, Wisconsin 54901

Annual Sale: They hold their annual Nordic ski sale in August. *Discount:* 10 percent, but only on package deals: Nordic ski packages; canoe or kayak boat and accessory packages; backpacks and additional supplies. *Discount to Nonprofit Organizations:* Varies. *Clearance Center:* They keep a basket near the door with markdowns. *Repairs:* Skis, fiberglass boat repairs. *Rent:* Sleeping bags, foam pads, tents, stoves, frame packs, soft packs, snowshoes, kayaks, canoes, cross-country skis. *Rental Sales:* Canoes are sold at the end of summer, skis in March, kayaks at any time. *R.P.:* If purchased within two weeks after rental.

WAUSAU

HIKE OUT, LTD.   715-842-0805
219 Jefferson Street
Wausau, Wisconsin 54401
   Annual Sale: They hold annual Nordic ski sales in August. *Discount:* They discount 10 percent, but only on package deals: Nordic ski packages; canoe or kayak boat and accessory packages; backpacks and additional supplies. *Discount to Nonprofit Organizations:* It varies. *Repairs:* Skis. *Rent:* Snowshoes, cross-country skis. *R.P.*

—————————————————————— **Wyoming**

CASPER

CROSS-COUNTRY MOUNTAINEERING   307-237-2071
128 West Second
Casper, Wyoming 82601
   Annual Sale: February. *Discount to Nonprofit Organizations:* 15 percent. *Repairs:* Minor repairs to skis and backpacking equipment. *Rent:* Sleeping bags, tents, frame packs, snowshoes, ice axes, cross-country skis. *Rental Sales:* Snow equipment is sold in March, other equipment in September. *R.P.*

MOUNTAIN SPORTS, INC.   307-266-1136
543 South Center
Casper, Wyoming 82601
   Rent: Sleeping bags, tents, frame packs, soft packs, snowshoes, downhill and cross-country skis, ice axes, crampons. *R.P.*

CHEYENNE

ALPINE HAUS SKI SHOP   307-635-2446
111 West 17th
Cheyenne, Wyoming 82001
   Annual Sale: Fall—ask them for this year's date. *Discount to Nonprofit Organizations:* 10 to 20 percent. *Repairs:* Everything

they sell. *Rent:* Sleeping bags, foam pads, tents, stoves, frame packs, downhill and cross-country skis.

CODY

SHOSHONE VALLEY TRADING COMPANY   307-587-9517
1340 Sheridan Avenue
Cody, Wyoming 82414
   Discount: 5 to 10 percent off list price on brands that may at times include Raichle, Kelty, North Face, Snow Lion, Vasque Voyageur, Galibier. *Manufacturers' Overruns:* 20 percent discount. *Repairs:* Bicycles, tents, bags, skis. *Rent:* Cross-country skis. *R.P.*

GILLETTE

DRIFTER SKI & SPORT   307-682-7922
Box 1426—Southview Shopping Center
Gillette, Wyoming 82716
   Annual Sale: Each spring. Ask them for this year's date. *Discount to Nonprofit Organizations:* 15 percent. *Repairs:* Ski repairs. *Rent:* Cross-country skis. *R.P.*

JACKSON

JACKSON HOLE SKI & SPORTS   307-733-4449
40 East Deloney Avenue—Box 2680
Jackson, Wyoming 83001
   Annual Sale: October. *Repairs:* Skis. *Rent:* Sleeping bags, foam pads, tents, frame packs, soft packs, snowshoes, downhill and cross-country skis, canoes. *Rental Sale:* October. *R.P.*

JACKSON SPORTING GOODS   307-733-3461
38 West Broadway—Box 468
Jackson, Wyoming 83001
   Annual Sale: First Saturday in November. *Discount:* Some merchandise is sold below manufacturers' list price. *Discount to Nonprofit Organizations:* 10 to 20 percent, but only to local schools and clubs. *Manufacturers' Seconds:* 20 to 40 percent off list price. *Repairs:* Skis, bindings, boots. *Rent:* Sleeping bags, foam pads, tents, stoves, frame packs, soft packs, downhill and cross-country

Approach to the Grand Teton
(Photo from the collection of North Country Mountaineering, Inc.)

skis, canoes, bicycles. *Rental Sales:* Mostly in November—at any time if there's something that interests you. *R.P.*

TETON MOUNTAINEERING   307-733-3595
Main Square
(P.O. Box 1533)
Jackson, Wyoming 83001
　Rent: Sleeping bags, foam pads, tents, stoves, frame packs, soft packs, snowshoes, kayaks, ice axes, crampons, cross-country skis. *Rental Sales:* Fall and spring—ask them for the dates.

WYOMING OUTFITTERS   307-733-3877
Box 1659—"On the Square"
Jackson, Wyoming 83001
　Rent: Snowshoes, downhill and cross-country skis. *R.P.*

LANDER

PAUL PETZOLDT WILDERNESS EQUIPMENT   307-332-4086
240 Lincoln
Lander, Wyoming 82520
　Semiannual Sales: April and November. *Discount:* There is about 10 percent discount on brands like Alpina, Kelty, Nordica. They carry their own excellent brand of equipment. *Discount to Non-profit Organizations:* 10 to 20 percent. *Manufacturers' Overruns and Seconds:* 20 percent savings—this is, of course, the best place to look for seconds of Petzoldt's own brand. *Clearance Center:* 20 to 40 percent savings. *Repairs:* Tents, sleeping bags, parkas. *Rent:* Cross-country skis. *Rental Sale:* April.

11

GORE-TEX® : SOMETIMES IT PAYS TO SPEND MORE

────────────────────── **Gore-Tex®: A Breakthrough**

In 1975 a major breakthrough in fabric for outdoor clothing was introduced. Manufactured by W. L. Gore and Associates, Inc., a medical supply house in Elkton, Maryland, this membrane was originally used during heart surgery to replace blood vessels. Because it is porous, yet impermeable to liquid, it allowed the blood to oxygenate without bleeding through. These same remarkable properties made Gore-Tex® a natural for the outdoor equipment market; it allows moisture, in the form of perspiration, to evaporate, yet it remains impermeable to rain and snow. The result is a fabric that is both breathable and waterproof.

The catch is that Gore-Tex® is still very expensive. But we think it's still a practical and worthy purchase for the budget-conscious; we've had the opportunity to field test Gore-Tex® wind shells, gaiters, an anorak, sleeping bags, and tent and have been thoroughly impressed by the results. In both light and pounding rain, in wind, and in below-zero temperatures, we've never worn anything as good. We sold our 60/40 parkas, gave away our Kelty windpants, finally retired our wind shirts, and almost totally switched to Gore-Tex.®

Eventually Gore-Tex® prices will probably come down, or someone else will figure out how to produce similar fabric. But in the meantime, Gore-Tex® alone allows backpackers and climbers the advantages of a lightweight, windproof, rainproof, yet ventile, breathing fabric.

According to the manufacturer of this "miracle" fabric, Gore-Tex® is a breakthrough in waterproof, comfortable outerwear technology. In developing a waterproof and yet comfortable outerwear material, the following six factors must be considered:

1. Waterproofness (water entry pressure)
2. Comfort (moisture vapor transmission rate)
3. Durability

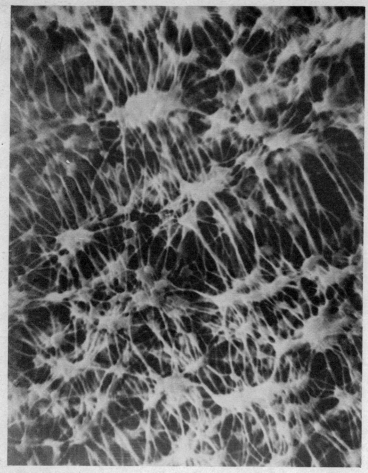

Scanning Electron Micrograph of Gore-Tex® Structure
(Photo courtesy of W. L. Gore & Associates)

4. Cleanability
5. Aesthetic properties
6. Cost and availability

The following technical information is excerpted from a bro-
chure published by W. L. Gore and Associates, Inc.

WATERPROOFNESS

There are two basic types of materials used in outerwear applications: water-repellent treated fabrics and fabrics coated with continuous polymeric films such as urethane. The repellents used on fabrics prevent the fabrics from being wetted, and, to a limited degree, prevent the penetration of the fabric by water. Typical tests of water repellency are the spray test and the static absorption test. Continuous coatings provide much greater resistance to water penetration than repellent treatments. Typical tests for coated fabrics are the impact penetration test, the rain test, and hydrostatic pressure tests.

The goal of the Gore-Tex® laminate work has been to provide a waterproof material with moisture vapor permeability sufficient to allow evaporative cooling and therefore comfort. The level of moisture vapor transmission required for the comfort of a resting subject is 350 to 600 grams of moisture vapor/square meter/24 hours. Under conditions of exertion and high environmental temperature, this requirement can exceed 2500 gm/M^2/24 hr., and there are conditions where comfort simply cannot be achieved. A realistic minimum goal for moisture vapor permeability is 2500 gm/M^2/24 hr. The data indicate that the moisture vapor permeability of Gore-Tex® laminates is essentially equivalent to a known comfort standard such as a repellent treated cotton/polyester woven fabric.

However, the real test of an outerwear material, especially one meant to be comfortable, is field evaluations. Human subject testing of Gore-Tex® laminates has been under way since 1974. In most instances, subjects with long experience in the performance of outerwear materials were selected to test garments in activities where they have expertise. Independent subjects as well as personnel from W. L. Gore and Associates, Inc., took part in the field testing. Some of the testing was performed by manufacturers of garments and tents. These activities included backpacking, hiking, mountaineering, sailing, canoeing, cross-country skiing, bicycling, and rock climbing. Items tested included rain suits, cagoules, parkas, tents, and jackets. In all cases the reports concluded that the Gore-Tex® laminates kept the testers dry under all the conditions tested and much more comfortable than any other waterproof material. In side-by-side comparisons with other waterproof materials, the testers wearing the other materials removed their garments, because of lack of comfort, and got wet in the rain, while the wearers of Gore-Tex® garments remained dry and comfortable.

Six subjects wearing rain suits of these materials were placed in an environmentally controlled room. The temperature was controlled at 120° F and the relative humidity at 20 percent. This condition was selected because it creates the maximum driving force for the escape of perspiration through clothing and is most effective in discerning the difference in moisture vapor transmission of materials. The subjects were to walk on a treadmill for 50 minutes, rest for 10 minutes, and repeat. The six subjects exchanged clothing materials so that all subjects evaluated all materials. The subjects' physiological conditions were continually monitored. The results of the tests were that only subjects wearing the repellent-treated cotton/-polyester (Quarpel) and those wearing Gore-Tex® laminate were able to walk for the full 50 minutes. Subjects wearing the other materials developed high skin and internal body temperature and high heartbeat rates and were in all cases removed from the test within the first 30 minutes.

In the field results to date, Gore-Tex® laminates have proven more comfortable than any other waterproof material tested.

COMFORT

Comfort is a highly subjective property and depends on the balance between an individual's thermal energy production and its exchange with his environment. Clothing modifies this balance. Physiologically an individual is comfortable if his skin temperature is 93°F $+$ 3° and the relative humidity at the skin surface is less than 100 percent.

DURABILITY

Durability is defined as how long the final garment will function as acceptable waterproof/comfortable outerwear. From this definition there are three ways the material can fail: It can wear out from use and abrasion; delamination can occur, making the garment unsightly; or the functionability of the Gore-Tex® layer can be impaired.

The overall wear resistance of the laminate is controlled by the selection of the fabric layers that go into the shell and liner. For high abrasion applications heavy, durable fabrics are chosen, and for light duty, lighter fabrics can be used.

Durability with respect to delamination can be a severe problem with all coated fabrics. Testing for long-term laminate quality, except by long-term field trials, has been difficult. A number of peel tests, flex tests, and laundering tests have been evaluated. Confidence in the validity of these tests is not high. Repeated home laundering cycles have been adopted as a standard delamination test with a minimum of five wash-dry cycles required as a screening evaluation. Correlation of this test with actual field use has not been completed, but field evaluation of current laminates has not shown any delamination problems.

Gore-Tex® laminates can lose their waterproof property in two ways: physical puncture or damage to the Gore-Tex® film in the laminate, and contamination of the Gore-Tex® by certain chemical agents. The former problem is common to any outerwear material and can be repaired by sewing, sealing or patching. The second problem is somewhat unique to Gore-Tex® and can be corrected by proper cleaning and rinsing.

CLEANING

The recommended cleaning technique for Gore-Tex® laminates is home laundering in a commercial low-sudsing detergent in cold or warm water. Agents that can contaminate Gore-Tex® and reduce waterproofness are various surfactants, some of which are found in high-sudsing detergents. Once contaminated by surfactants, waterproofness can be regained by properly rinsing these surfactants out of the Gore-Tex®.

Dry cleaning is not recommended for Gore-Tex® laminates because of the high surfactant concentration found in commercial dry cleaning solutions. Pure dry cleaning solvents do not affect the laminates and, in fact, are most successful in removing harmful surfactant contamination from the Gore-Tex® pores.

AESTHETIC PROPERTIES

Included in this category are color, texture, and laminate "hand."

Color and texture, like mechanical durability, are functions of shell and liner fabrics and can be controlled by proper selection. "Hand," suppleness, and rustle are all functions of fabric selection and laminating process. No attempt has been made to qualify this property, but subjective comparisons have been made. Gore-Tex® laminates from a qualitative standpoint have aesthetic characteristics comparable to most coated fabrics.

COST AND AVAILABILITY

After four years of research, Gore-Tex® laminates were made commercially available as outerwear materials on January 1, 1976. These laminates are manufactured and marketed solely by W. L. Gore and Associates, Inc., Elkton, Maryland. Current pricing of these laminates is dependent upon construction.

Pessimism concerning the possibility of producing a waterproof, breathable material has been widespread. Written opinions that the combination of these two properties in a single material is unobtainable are numerous. Gore-Tex® laminates now make the seemingly impossible possible.

A GORE-TEX® SAMPLER: WHO'S PRODUCING WHAT

A-16 WILDERNESS CAMPING OUTFITTERS   714-283-2374
4620 Alvarado Canyon Road
San Diego, California 92120

A-16 is producing Gore-Tex® rainjackets. We haven't had the opportunity to test any of them, but A-16 products are generally of very high quality.

ALTI-PRODUCTS   505-982-5065
129 West Water Street
Santa Fe, New Mexico 98501

ALTI-Products is now producing a jacket called the Cape Parka. Only the shoulders and hood are made of Gore-Tex®. The rest of the parka is constructed of Sierra cloth. This parka costs about as much as an all Gore-Tex® parka, so we don't think it really makes sense to buy it.

ARCATA TRANSIT AUTHORITY   707-822-2204
650 Tenth Street
Arcata, California 95521

The best Gore-Tex® item made by this company is a sleeping bag cover called the Bear Necessity. It's offered in two fabric combinations. In both versions, the top fabric is Gore-Tex® film laminated between 1.5 ounce ripstop nylon and 0.8 ounce nonwoven polyester. The less expensive version has a floor of 2.2-ounce nylon taffeta coated with Super K-Kote, the most durable waterproof coating available on lightweight fabric. In moderate use, and with reasonable care, this fabric will remain waterproof for a number of years. But, as with any coated fabric, the waterproofing will gradually wear off.

A more expensive version of the Bear Necessity features a floor material of Gore-Tex® film laminated between 1.9-ounce nylon taffeta and 1.5-ounce nylon tricot knit. This fabric adds 4 ounces to the weight of the bag cover, but it will not lose its ability to repel water. You would have to wear through the outer protective fabric before you could permanently damage the waterproof layer of the Gore-Tex® film.

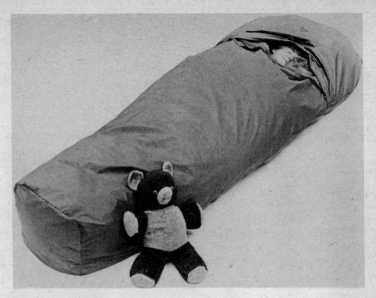

The Bear Necessity
(Photo courtesy of Arcata Transit Authority)

BANANA EQUIPMENT COMPANY   303-585-5824
351 Moraine Street
Estes Park, Colorado 80517

Each product made at Banana is sewn—from start to finish—by a single seamstress, to maintain pride in the workmanship—and to prevent insanity. Obviously this is working out just fine; their quality is consistently excellent.

Banana Equipment is the only producer right now of Gore-Tex® bibs. These are terrific garments for ice climbing, skiing, and winter hiking. The bibs have two large pockets, and a flap-covered 22-inch coil zipper that provides for ventilation and easy removal. Be sure to order your pair with bottom zippers so you don't have to take your boots off to get them on.

Banana's Zip-Front Jacket is one of the best-designed jackets we have seen. The fit is excellent and allows unrestricted arm movement even when worn over a bulky parka. The hood will fit over your helmet without blocking your vision. It's a splendid parka and good looking, too.

Banana Equipment Gore-Tex® Bib
(Photo courtesy of Banana Equipment)

THE BOULDER MOUNTAINEER

303-442-8355

1329 Broadway
Boulder, Colorado 80303

Bob Culp, one of the most innovative and dynamic climbers of the early 1960s, is producing rainsuits, climbing overalls, and bivouac bags of Gore-Tex® fabric. These items are available on a rental basis as well as for sale at the Boulder Mountaineer retail store. We've seen some of the prototypes and are impressed by the care the seamstresses who work for Culp have taken. All of his Gore-Tex® equipment is competitively priced, and they will custom-construct any garment you want of Gore-Tex®.

CLEAR CREEK TENTS

415-351-9763

14361 Cataline Street
San Leandro, California 94577

Clear Creek produces a Gore-Tex® tent called the Ogive. It's a free-standing tent that has no guylines or tent pegs. Because the support mechanisms for this tent are on the outside, it's very roomy and comfortable. It's also very stable.

EAGLE PEAK

916-784-0600

Box 1294
323 Lincoln Avenue
Roseville, California 95678

Eagle Peak is producing Polarguard sleeping bags with Gore-Tex® shells. These are distributed by Feathered Friends in Seattle, Washington.

EARLY WINTERS, LTD.

206-633-5203

110 Prefontaine Place South
Seattle, Washington 98104

Early Winters, Ltd. was one of the first companies to produce Gore-Tex® gear. Their gear is still among the very best.

With the exception of Rivendell tents, Early Winters tents are the most superior shelters available in North America. The Light Dimension, at 3½ pounds, is a sleek, elegant, two-person backpacking tent that can be set up in one minute. It requires only three pegs. This tent has a large mosquito-net-covered vent in the back that allows free flow of air. It uses very lightweight shock-cord loaded poles—the same poles used in the Early Winters Omnipotent. The

**Banana Equipment Zip-Front
Gore-Tex® Jacket**
(Photo courtesy of Banana Equipment)

**Boulder Mountaineer
Gore-Tex® Rain Parka**
(Photo courtesy of
the Boulder Mountaineer)

semi-circular profile of the tent gives the backpacker more than 50 percent more usable space than an A-frame tent of the same height and width. The door has the finest insect netting we've seen on any tent; it will even stop "no-see-ems." We field-tested ours during a week of intense Vermont rain. All seams had previously been sealed three times and had absolutely no leakage. We developed no condensation problem.

Early Winters produces a Gore-Tex® parka and a Gore-Tex® pullover anorak. The parka has zippered underarms, handwarmer pockets, two very large cargo pockets that seal with Velcro, and a hood with a jaunty visor front. The material used is 1.9-ounce nylon taffeta laminated over a Gore-Tex® membrane. The parka weighs 17 ounces, and is built like iron. The anorak weighs 15 ounces.

Their Bivy Sac is the simplest of all bivouac covers. It's designed to be used with or without a sleeping bag. It's extremely lightweight and should be included in your pack as an emergency shelter.

ELD EQUIPMENT   206-357-4812
117 North Washington
Olympia, Washington 98501

The most extraordinary piece of equipment produced by Eld Equipment is also the most useful for backpackers. It's an Over-Pack Rain Parka of Gore-Tex® that gives complete protection from wind-driven rain. This thigh-length parka is large enough to cover the person *and* the pack. Instead of the open sides of a poncho, this parka has sleeves with Velcro closed cuffs, and a full-length Talon #6 two-way coil zipper down the front. There is a flap over the zipper held closed with Velcro spots.

Because it goes over the pack, this parka gives better ventilation, reduces condensation, and allows the wearer to pull his or her arms inside for adjustment of the pack straps, access to pockets, etc. It can also be used without a pack as a general purpose raincoat.

Of interest to the expeditionary climber is the fact that Eld will custom-construct Gore-Tex® box tents. To date, a prototype has been used on Mount Foraker in Alaska and has stood up well to severe weather conditions.

FEATHERED FRIENDS   206-632-4402
1314 N.E. 43rd Street
Seattle, Washington 98105

Feathered Friends is a remarkable small firm which has blended

Early Winters Gore-Tex® Light Dimension Tent
(Photo courtesy of Early Winters, Ltd.)

together the best ingredients to produce what must be one of the finest white Polish goose down Gore-Tex® bags. White Polish down provides 30 percent more warmth for the weight of the bag. The down pod's longevity is greater than that of any other down bag. Initially, the Gore-Tex® bag with Polish down is a larger investment than a Gore-Tex® bag using Mainland China down. But we consider these bags well worth the investment. Their warmest bag is the Snowy Owl Expedition Mummy; it's much too warm a bag even for the winter backpacker, but it ought to be considered by expeditionary climbers. Our favorite (and more practical) bag is the Tern. It weighs only 3 pounds 3 ounces, and has 8 inches of loft.

FORREST MOUNTAINEERING   303-433-3373
1517 Platte Street
Denver, Colorado 70202

Bill Forrest produces two very interesting Gore-Tex® items. Winteralls are a multipurpose winter overall designed for ice climbing, snowshoeing, and winter camping. They are easy to get on and off: There is a full front opening with a double-pull zipper and a snapped flap; also the pants legs have zippers so you can put them on over double boots. These overalls are designed to accommodate layers of warm winter clothing. Beneath the rear pocket is a zippered "outhouse opening." These overalls come unhemmed so you can finish them to the right length for you. Various closure options for the cuff, such as buckle-strap combinations or draw cords, may be installed, though most people will cover the cuff with a gaiter.

Another interesting Forrest Gore-Tex® item is the full-length bivouac sac—big enough to provide protection without feeling cramped or constricted while wearing high loft clothing. The Full Length Sac has a zipper from the waist to the built-in hood. To give maximum weather protection, these sacs are made of coated nylon; therefore it's advisable to leave them open during fair weather to avoid condensation. For maximum waterproofness, the seams should be sealed by the user.

HOLUBAR MOUNTAINEERING   303-442-7656
6278 Arapaho Avenue East
Boulder, Colorado 80302

Holubar at this time is producing only one Gore-Tex® item, the Holubar Gore-Tex® Mountain Parka. It is styled after the traditional 60/40. We think it has too much stitching over the entire body

Gore-Tex® Astro Overboots
(Photo courtesy of Forrest Mountaineering)

and consequently requires too much waterproofing. It is, however, extremely handsome and definitely suited for country club wear.

KELTY PACK COMPANY   213-768-1922
9281 Borden Street
Sun Valley, California 91352
 Kelty is now producing a full Gore-Tex® suit made of 1.9 ripstop nylon outside, with a 1.5 tricot inner lining. At the time of this writing we have not field tested this Gore-Tex® suit, but it looks strong, sturdy, and extra-expensive—like most Kelty products.

KEN'S MOUNTAINEERING   702-786-4824
155 North Edison Way
Reno, Nevada 89502

Ken's Mountaineering is a custom outfitter and has designed and manufactured products for expeditions to the Himalayas, the Andes, all major North American ranges, and Antarctica and Greenland. Here you can order anything you want of Gore-Tex®.

INTERNATIONAL SPORTSWEAR   802-334-6722
Longview Street
Newport, Vermont 05855

International Sportswear makes full Gore-Tex® suits that are used in international sailing competitions. The overalls are nicely styled, with bell bottom legs that you can tighten with Velcro. These overalls are easily adaptable for backpacking use.

LOG HOUSE DESIGNS   603-694-3183
Chatham, New Hampshire 04058

Log House produces a Gore-Tex® parka that's very simple in design, and relatively inexpensive for a Gore-Tex® item. Their best Gore-Tex® product is the gaiter. It closes with a YKK Delrin zipper and has a nose hook. Here you can find very fairly priced products.

LOWE ALPINE SYSTEMS   303-665-9220
1752 North 55 Street
Boulder, Colorado 80301

The Lowe Makalu Gore-Tex® Dome Tent is the most advanced dome tent being produced in the United States. Until now it has been mandatory in dome designs to use poles in external sleeves since the sleeves helped to space the fly away from the main tent body. Gore-Tex® laminates have eliminated this restriction, and the Makalu Domes have all of their pole sections inside the tent, so users can get in out of the weather and set up the tent entirely from the inside. An additional advantage of this system is that when the pole structure flexes under wind loads the entire tent tightens up; it thus has much greater free-standing strength. Since the poles do not have to be threaded into sleeves, setup time is substantially reduced. Once the tent is erected, a series of short web straps are snapped around the pole sections to hold them in place during high wind conditions.

For go-light backpackers and climbers who may not find room to set up a larger tent, the Makalu II is a snug little tent that will sleep two, yet offers more interior volume than other tents with similar floor dimensions. Because of its smaller size and the increased strength due to the inside poles, it is virtually unshaken by severe

Lowe Gore-Tex® Dome Tents
(Photo courtesy of Lowe Alpine Systems)

winds and easily sheds heavy snow loads. Designed for and tested on the recent International Makalu Expedition, this tent is already in demand by climbers doing major winter ascents using alpine style techniques rather than fixed camps. The matching doors at either end offer easy access and are backed by mosquito netting for improved ventilation. There are two inside pockets for storage of small items.

Virtually identical to the Makalu II, except for its larger dimensions, the Makalu IV provides spacious accommodations for larger groups. Four six-foot adults in mummy bags can sleep on the floor, and there is sit-up space for twice that many. For two or three people, it provides ample room for the occupants and all their gear. Getting dressed is much simpler, since the 59-inch apex allows partial standing instead of squirming around on the floor trying to put on a pair of pants. The large double doors provide excellent ventilation on hot nights when mosquitoes become your primary enemy. Two inside pockets are included.

MARMOT MOUNTAIN WORKS   303-243-2339
226 Litkin Avenue
P.O. Box 2433
Grand Junction, Colorado 81501

Marmot produces the finest down clothing available with Gore-Tex® shells. All their garments contain white Polish goose down with a minimum lofting capability of 650 cubic inches per

Marmot Mountain Works Gore-Tex® Sleeping Bags
(Photo courtesy of Marmot Mountain Works)

ounce. Most of their down lofts as high as 750 cubic inches per ounce. White Polish down has 20 percent more lofting ability than the highest-graded down currently being used by 90 percent of the outer gear industry. At Marmot Mountain Works a single person sews and constructs an entire garment. Each individual piece of nylon is hot cut by hand to insure that there will be no unraveling. The fabric chosen for the outside of all Marmot's down garments is 2-ounce nylon.

The Marmot Gore-Tex® Sweater is filled with 8½ ounces of goose down and has a draft tube behind the zipper. It has down-filled handwarmer pockets. It's an excellent all-purpose lightweight down garment.

The Golden Mantle Parka truly represents the state of the art in down gear. This is the lightest weight expedition parka in its class. With 1 pound of down insulation, it's total weight is 2½ pounds. But you don't really need a parka this warm unless you're backpacking in the severest conditions.

Marmot's sleeping bags are the best-designed and warmest sleeping bags available in the United States. Marmot uses a differential cut design that assures even loft. A double draft tube guarantees a windproof seam behind the zipper. The foot section of their

Marmot Mountain Works Golden Mantle Parka
(Photo courtesy of Marmot Mountain Works)

Marmot Mountain Works Down Sweater
(Photo courtesy of Marmot Mountain Works)

bag has seven separate down-filled compartments. Most other manufacturers use only four. The loft around the foot area is dramatically higher than any other sleeping bag we've seen.

Marmot's Burrow is a one-person shelter designed for the solo backpacker and for alpine bivouacs. The top of the shelter is of Gore-Tex® laminated fabric. Three poles form the hood section. The floor is of urethane-coated nylon. This 1-pound-13-ounce sweater will increase the warmth of your sleeping bag by 10 to 20 degrees.

MOONSTONE MOUNTAINEERING   707-822-9471
1021 H Street
Arcata, California 95521

Fred Williams at Moonstone Mountaineering is producing Gore-Tex® garments as well as sleeping bags. Moonstone Mountaineering was the first manufacturer of sleeping bags to combine Polarguard with Gore-Tex® shells. If you are considering a sleeping bag to take on kayak or canoe trips, or if you are going someplace where there will be rainy or snowy conditions for some time, this combination is really excellent. Anna owns a Moonstone bag, and it has served her well. Moonstone is a small manufacturer and so seems to care about the quality of its merchandise. We've found it to be first-rate.

MOUNTAIN SAFETY RESEARCH   206-762-4244
631 South 96th Street
Seattle, Washington 98108

Mountain Safety Research produces three different styles of Gore-Tex® tents: a dome tent, a cone tent, and a quonset tent.

The MSR tents have yellow 1.1-ounce nylon ripstop on the weather side for strength and abrasion resistance, and white non-woven polyester on the inside for condensing and capturing water vapor, passing it through the Gore-Tex® membrane to the outside. This Gore-Tex® tent offers the user freedom from condensation problems in conditions above 28°F.

The two entrances have double overlapping silicone-treated zippers to prevent rain leaks. A hood over the zip door allows for ventilation when raining. All openings have insect netting. The floors are made of 2.8-ounce coated nylon taffeta and extend 6 inches up the sides.

Poles or hoops are made of curved tubular sections joined with inside shock cords. They fold into a compact bundle about 17 inches long. MSR has poles ⅜ and ⁷⁄₁₆ inches in diameter. If you are a three-season backpacker and you take care not to let snow accumulate to a thickness of more than one inch, the lighter weight poles will serve just fine; if you do lots of winter camping, the heavier poles are the better choice.

The MSR Gore-Tex® Cone Tent is a lightweight tent that has plenty of room, more room at the foot than similar designs by other manufacturers. Support is provided by two tubular aluminum hoops. A small rear cone of coated nylon has no floor, and can be used for gear storage. The cone is separated from the living space by zip panels of nylon cloth, and insect netting. The main compartment is 101 inches long; width is 53 inches at head end and 36 inches at the foot. The head height is 40 inches; the foot is 26 inches.

MOUNTAIN TRADERS   415-845-8600
1700 Grove Street
Berkeley, California 94709

Mountain Traders produces a nicely designed line of Gore-Tex® gaiters for the hiker, backpacker, mountaineer, and ski tourer. They come in heights of 6 to 16 inches with Velcro and zipper closures.

OUTDOOR GEAR
Box 369
Centre Harbor, New Hampshire 03226

At present, Outdoor Gear is manufacturing two Gore-Tex® garments: the Nutshell and the Roo. In the design stage are mitten shells, rain pants, full front-zippered parkas, and packs.

The Nutshell and Roo are simple functional anoraks made from the same basic pattern. The Nutshell is available in lightweight urethane coated nylon or a Gore-Tex® ripstop laminate. It can be stuffed into its own pocket and carried on an equipment sling. It has a side zipper for ventilation and accessibility.

The Roo is made of heavier 1.9-ounce taffeta Gore-Tex® laminate and has a front zipper. It has a large "kangaroo" pocket. It does not, however, stuff into its own pocket.

Both anoraks sell for under $50.

POWDERHORN MOUNTAINEERING   307-733-3365
Box 166
Jackson, Wyoming 83001

We have been told by the people at Powderhorn that they plan to produce a Gore-Tex® version of the Wind River parka. We've always liked the design of the Wind River parka; it's both practical and very stylish. When this new parka does hit the market, we're sure it will warrant your consideration.

RECREATIONAL EQUIPMENT
INCORPORATED   800-426-4840 (toll free)
1525 11th Avenue
Seattle, Washington 98134

We're very disappointed that Recreational Equipment, Inc., has chosen not to give its customers any real break on the price of Gore-Tex® gear. A very similar parka to the one it sells for about $65 sells for at least $10 less at both Outdoor Gear, P.O. Box 369, Centre Harbor, New Hampshire 03226, and International Mountain Equipment, in North Woodstock, New Hampshire. The Gore-Tex® anorak featured by REI is about $15 more expensive than very similar garments offered by Outdoor Wear and International Mountain Equipment.

RIVENDELL MOUNTAIN WORKS   208-787-2746
Box 198
Victor, Idaho 83455

We have been using Rivendell products for the past six years. They have excellent standards of quality control. We think every item they produce is fantastic.

Rivendell's Gore-Tex® Bombshelter tent is the strongest two-person A-frame tent available. The Bombshelter combines all the advantages of the St. Elias tent by North Face, the Sierra Designs' Glacier, and the Gerry Expedition Mountaineering tent.

Rivendell's Hot Tamale Shell is the most functional and best-made mitten shell we know. The shell's palm and thumb are of 8-ounce double-coated nylon duck. The palm and inside of the thumb are covered with a dense smooth-skinned neoprene. This acts as a great insulator, is abrasion-resistant, provides a nonslip grip, and won't allow ice buildup. The back is nylon oxford laminated to Gore-Tex® film and nylon tricot knit, so wet hands don't become a problem, either from sweating hands, rain, or melting ice.

Gore-Tex® Bombshelter Tent
(Photo courtesy of Rivendell Mountain Works)

The nonslip 65/35 cloth lining is comfortable against the skin for those who have occasion to use the shell without a mitten. The sleeve, which is long and large in circumference, closes with a simple draw-string toggle, easily operated with one hand and your teeth.

Rivendell's new Elf Boot adds comfort to the advantages of light-weight Nordic ski touring equipment. The fact that it is an overboot —enclosing the entire leg, shoe to knee—offers warmth and security from the elements unattainable by gaiters.

The foot section is a high quality vinyl-coated polyester. It is completely waterproof, very strong, and flexible to −50°F. It is lined with a thin layer of foam-backed nylon tricot. This provides the insulation necessary for warm feet during lunch and rest stops. A heavy leather square on the bottom protects the Elf Boot from your heel plate. From the ankle up, nylon oxford is laminated to Gore-Tex® with a tricot backing. The top and instep are drawn up by means of a simple toggle.

Elf Boots can be used with some cable bindings. But they were designed specifically for use with pin bindings and will fit over ski touring shoes only. In camp, Elf Boots make splendid mukluks over down booties.

Rivendell's new Gore-Tex® Cagoule is a simply cut garment with a skirt large enough to allow the knees to be drawn inside during a downpour. The hood design is ample enough to accommodate a climbing helmet, and the stiffened visor prevents water from dripping on your nose while affording unrestricted visibility. The throat opens by means of a protected zipper. There are no shoulder seams. The large interior pocket is covered by a generous weather flap, and is closed by a coil zipper with a pull slider. Velcro adjustments close the cuffs and the skirt may be cinched up around the waist.

The Rivendell Anorak is cut shorter and slimmer than their cagoule, but it is identical in other respects. If you combine the Anorak with Rivendell Rain Pants, you'll have formidable protection and mobility in the fiercest storms. This is an ideal combination for outdoor activities in marginal weather, and opens up a new world to backpackers venturing into foul weather mountains.

THE SESAME SNAP   206-733-0123
100 East Pine
Bellingham, Washington 98225
    Sesame Snap produces a very simply designed Gore-Tex® rainjacket. Very little stitching makes this one of the most waterproof Gore-Tex® garments.

SIERRA WEST   805-963-8727
6 East Yanonali Street
Santa Barbara, California 93101
    Sierra West is currently producing more Gore-Tex® shell garments than any other manufacturer. Their equipment is available at about four hundred stores across the United States. They produce Gore-Tex® parkas, pants, an anorak, and a cagoule all under their own label.

SYNERGY WORKS   209-233-5213
255 Fourth Street
Oakland, California 94607
    Synergy produces the most expensive Gore-Tex® parka manufactured in the United States (slightly under $150). The budget-minded shopper can buy a parka that will serve just as well for a smaller investment.

WESTERN MOUNTAINEERING   408-292-4445
550 South First Street
San Jose, California 95113

Sierra West Gore-Tex® Storm King Parka
(Photo courtesy of Sierra West)

Western Mountaineering is making some very nice down sleeping bags that are covered with Gore-Tex®; among them are the Middlebag and the Narrowlite.

The Western Mountaineering Middlebag is a very popular bag featuring sophisticated construction and a generously cut mummy shape. Sensible proportions make it comfortable, not claustrophobic. It is cut large enough to accommodate extra clothing, boots, waterbottle, or anything else you might want to keep from freezing on a cold night. It comes in four different lengths and two different weights of down filling.

The Narrowlite is a lightweight, compact mummy bag designed primarily for the backpacker and mountaineer. It is form-fitting, offering the greatest warmth-to-weight efficiency in their line.

THE DRAWBACKS OF GORE-TEX®

The main drawback of Gore-Tex® is the absolute necessity to seal all stitched seams. Most companies recommend K-Kote sealant. Our personal experience with this seam sealant on Gore-Tex® fabric has been unfavorable. We used three individual applications of K-Kote on our parkas. The seams soon started to leak anyway. We realized the leakage was due to cracks in the water proofed areas caused by stuffing the garments into stuff sacks. Even with great care, the longevity of a K-Kote seal on Gore-Tex® is too brief. K-Kote sealant does apply very easily, and we've used it with great success on conventional fabrics.

Marmot Mountain Works, Feathered Friends, and Early Winters Ltd. are all recommending Rain-Coat sealant, which is distributed by Early Winters Ltd. We tried this seam sealant on our Gore-Tex® tents, bib overalls, and bivouac bags. After three individual applications, spread over a two-day period to allow for drying, we have had no leakage at the seam after using this sealant. So we feel confident about recommending it for use with Gore-Tex®. However, beware, for this sealant is *highly toxic*. Steve made the mistake of using it in a room that had only one window open, and wound up feeling nauseated and suffering a pounding headache. Another problem with Rain-Coat is that it must be applied with a brush, so you must take care not to let it get too thick.

W. L. Gore produces its own sealant, called Seam-Stuff. Outdoor Gear in Centre Harbor, New Hampshire, has been experimenting with the sealant. The results have been mixed. Mountain Safety Research has informed us that they are not yet ready to recommend this particular sealant. We used Seam-Stuff on only one of our garments; afterward it did not leak. If additional reports confirm the success of this sealant, it may be worth your while to consider using it.

No matter what sealant you use, do not let it touch your skin. Also, do not rush the sealing process. Make sure there's adequate time for each coat to dry. We recommend three coats for each seam.

WHERE TO BUY GORE-TEX® BY THE YARD

It goes without saying that you can radically decrease the cost of equipping yourself with Gore-Tex® gear if you make it yourself. The following retail outlets and manufacturers sell Gore-Tex® fabric for home sewing.

Albany, Inc.
3040-B Eastway Drive
Charlotte, North Carolina 28205

Altra, Inc.
5441 Western Avenue
Boulder, Colorado 80301

Barons Fabric
22914 Victory Boulevard
Woodland Hills. California 91367

Early Winters, Ltd.
300 Queen Anne Street N.
Seattle, Washington 98101

Eastern Mountain Sports, Inc.
1041 Commonwealth Avenue
Boston, Massachusetts 02215 (And all EMS retail outlets)

Marmot Mountain Works, Ltd.
3049 Adeline Street
Berkeley, California 94703

Marmot Mountain Works, Ltd.
546 Main Street
Grand Junction, Colorado 81501

Nippenose Equipment Company
225 West Fourth Street
Williamsport, Pennsylvania 17701

Recreational Equipment Inc.
San Pablo Avenue
Berkeley, California 94702

Recreational Equipment Inc.
1525 11th Avenue E
Seattle, Washington 98134

The Ski Hut
1615 University
Berkeley, California 94703

Wilderness Ways
P.O. Box 961
146 East Main Street
Newark, Delaware 19711

WHERE TO BUY OUTDOOR EQUIPMENT KITS OF GORE-TEX®

Country Ways, Inc.
3500 Highway 101 South
Minnetonka, Minnesota 55343

APPENDICES

———————— **I. Sources of Outdoor Equipment Kits**

With only two exceptions, each of the outdoor equipment kit manufacturers listed below publishes a mail-order catalog. Unless we
have added additional comments after the address, each produces
a standard range of outdoor gear kits: rain pants, parkas, sleeping
bags, vests, soft packs. These kits will save you about 50 percent
of the cost of buying pre-sewn gear.

ALTRA, INC.   303-449-2401
5441 Western Avenue
Boulder, Colorado 80301

BLACK FOREST ENTERPRISES   916-265-4845
Box 1007
Nevada City, California 95959
   Black Forest produces two snowshoe kits: the Freestyle, and the
Travel-light. These are easy to construct and can carry loads up to
500 pounds.

CALICO KITS   1-800-525-3701
1275 Sherman Drive
Longmont, Colorado 80501
   In addition to the usual range of gear kits, Calico produces kits
for chamois shirts.

COUNTRY WAYS   612-473-4334
3500 Highway 101 South
Minnetonka, Minnesota 55343
   In addition to the usual gear, Country Ways offers custom cutlery
kits (steak knives, camp knives, boning knives, etc.). They also sell
kits for cross-country skis, frame looms, floor looms, hearth brushes,

thumb pianos, banjoes, dulcimers, canoe paddles, and canoe and kayak roof rack brackets.

EASTERN MOUNTAIN SPORTS KITS   603-924-7276
Vose Farm Road
Peterborough, New Hampshire 03458

FROSTLINE KITS   303-451-5600
Frostline Circle
Denver, Colorado 80241
    Frostline was the originator of the outdoor equipment kit idea and has a fabulous range of kits, including down pillows and comforters, panniers, garment carriers, tote bags, and ski suits.

MOUNTAIN SEWN PATTERNS   1-800-426-4840 (except in
Recreational Equipment, Inc.   Washington, Alaska, Hawaii)
P.O. Box C88-125   In Washington:
Seattle, Washington 98188   1-800-562-4892
                   Alaska, Hawaii, toll calls:
                   206-575-4480
    This is the only place you can buy just equipment patterns without pre-cut fabric and fill.

PLAIN BROWN WRAPPER, INC.
2055 West Amherst Avenue
Englewood, Colorado 80110
    In addition to the usual range of gear kits, they produce a kit for bib overalls.

ROYAL DOWN PRODUCTS, INC.
101 North Front Street
Belding, Michigan 48809
    Royal Down doesn't publish a mail-order catalog. If you write to them they'll supply you with a list of retail outlets where you can purchase their kits.

SEW IT YOURSELF KITS
Holubar Mountaineering
Box 7
Boulder, Colorado 80306

1-800-525-2540
303-443-8403
in Colorado, Alaska, Hawaii

In addition to the usual range of outdoor equipment kits, Sew It Yourself Kits produces kits for Western-style chamois shirts, cargo and tote bags, garment bags, ski bibs, ski jackets, and insulated mittens. They also have kits that will allow you to personalize your gear.

SHERPA DESIGNS, INC.
3109 Brookdale Road East
Tacoma, Washington 98446

206-531-4114

Sherpa Designs produces the finest snowshoe kits on the market. Their directions are extremely easy to follow.

SUTTER'S MILL
810½ Main Avenue
Moorhead, Minnesota 56560

218-233-8990

This company is so new that they don't yet publish a mail order catalog. In addition to the usual gear kits, they sell kits for moccasins and mukluks. They've just begun to offer kit sewing courses and courses in custom designing your own equipment.

TUBBS
Wallingford, Vermont 05773

They sell traditional snowshoe kits with catgut lacing, very good kits that come in several different styles.

————II. Summer Day Hike Equipment Checklist

—— daypack
—— poncho, or rainjacket and pants
—— hiking shoes
—— extra socks
—— bandana
—— hat with sunvisor
—— canteen
—— repair kit (needle and thread, safety pin, ripstop nylon repair tape)
—— flashlight with spare batteries and spare bulb
—— map and compass

- —— warm shirt, or sweater
- —— pocketknife
- —— small first aid kit, including moleskin and compact scissors
- —— metal mirror
- —— sunglasses
- —— suntan cream
- —— Cutter insect repellent
- —— matches in waterproof case, or waterproof matches
- —— food for the day
- —— plastic bag for packing out debris
- —— 5 or 6 dimes for emergency phone calls
- —— toilet paper

—— III. Summer Weekend Backpack Checklist

- —— frame pack and bag, or internal frame pack
- —— Ensolite pad
- —— sleeping bag in waterproof stuff sack
- —— waterproof groundsheet
- —— tent
- —— stove
- —— fuel in fuel bottle
- —— cook kit
- —— spoon, or fork, knife, and spoon
- —— cup and plate
- —— poncho, or rainjacket and pants
- —— hiking shoes
- —— three extra pairs of socks
- —— extra underwear
- —— bandana
- —— hat with sunvisor
- —— canteen
- —— repair kit (needle and thread, safety pin, ripstop nylon repair kit)
- —— flashlight, with extra batteries and extra bulb
- —— map and compass
- —— warm shirt
- —— down or synthetic-filled vest (or heavy sweater)
- —— pocketknife
- —— first aid kit, including moleskin and scissors
- —— toilet kit (toothbrush and tooth powder or paste, metal mir-

ror, comb, toilet paper and *biodegradable soap.* You can buy
biodegradable soap in a health food store.)
—— plastic or canvas washbasin
—— sunglasses
—— suntan cream
—— Cutter insect repellent
—— matches in waterproof case, or waterproof matches
—— food
—— food bag
—— couple of plastic bags for packing out debris
—— extra loose-fitting clothing
—— wristwatch (if you wish)
—— 6 or 8 dimes for emergency phone calls (we recommend so
many because they tend to get lost)